Biased Embryos and Evolution

What determines the direction of evolutionary change?

This book provides a revolutionary answer to this question. Many biologists, from Darwin's day to our own, have been satisfied with the answer, 'natural selection'. Professor Wallace Arthur is not. He takes the controversial view that biases in the ways that embryos can be altered are just as important as natural selection in determining the directions that evolution has taken, including the one that led to the origin of humans. This argument forms the core of the book. However, in addition, the book summarizes other important issues relating to how embryonic, and post-embryonic, development evolves.

Written in an easy, conversational style, this is the first book for students and the general reader that provides an account of the exciting new field of evolutionary developmental biology ('evo-devo' to its proponents).

WALLACE ARTHUR is in the Department of Zoology at the National University of Ireland, Galway. He is the author of six previous books.

Biased Embryos and Evolution

WALLACE ARTHUR

Professor of Zoology
National University of Ireland, Galway

PUBLISHED BY THE PRESS SYNDICATE OF THE UNIVERSITY OF CAMBRIDGE
The Pitt Building, Trumpington Street, Cambridge, United Kingdom

CAMBRIDGE UNIVERSITY PRESS
The Edinburgh Building, Cambridge CB2 2RU, UK
40 West 20th Street, New York, NY 10011–4211, USA
477 Williamstown Road, Port Melbourne, VIC 3207, Australia
Ruiz de Alarcón 13, 28014 Madrid, Spain
Dock House, The Waterfront, Cape Town 8001, South Africa

http://www.cambridge.org

First published 2004

Printed in the United Kingdom at the University Press, Cambridge

Typefaces Trump Mediaeval 9.5/15 pt. and Times   *System* LaTeX 2_ε [TB]

A catalogue record for this book is available from the British Library

Library of Congress Cataloguing in Publication data
Arthur, Wallace.
Biased embryos and evolution / by Wallace Arthur.
    p.   cm.
Includes bibliographical references (p.) and index.
ISBN 0 521 83382 5 – ISBN 0 521 54161 1 (pbk.)
1. Developmental biology.   2. Embryos.   3. Evolution (Biology)   I. Title.
QH491.A768   2003
571.8 – dc22   2003062621

ISBN 0 521 83382 5 hardback
ISBN 0 521 54161 1 paperback

Stephen Jay Gould once said that *the most noble word of all human speech* is *'teacher'*. In the spirit of that remark, I dedicate this book to all of my teachers across the years, and in particular: my mother, my father (*in memoriam*), Amyan Macfadyen, Bryan Clarke, Alec Panchen, and, during an all-too-brief period of sabbatical leave spent at Harvard in 1987, SJG himself. I thank you all.

Contents

Preface   *page* ix

Acknowledgements   xii

1 The microscopic horse   1

2 What steers evolution?   9

3 Darwin: pluralism with a single core   26

4 How to build a body   40

5 A brief history of the last billion years   54

6 Preamble to the quiet revolution   67

7 The return of the organism   76

8 Possible creatures   88

9 The beginnings of bias   105

10 A deceptively simple question   117

11 Development's twin arrows   128

12 Action and reaction   140

13 Evolvability: organisms in bits   152

14 Back to the trees   159

15 Stripes and spots   175

16 Towards 'the inclusive synthesis'   191

17 Social creatures   201

Glossary   211

References   223

Index   231

Preface

The processes of life work at many different speeds. Some occur swiftly, others require millions of years; some recur in a cyclical manner, others are historically unique. The two great processes of biological creation are embryological development and organic evolution. Development fits into the rapidly recurring category, while evolution fits into the complementary category of the unique and time-extended.

However, in the world of life, always beware of generalizations. It is science's duty, and its pleasure, to attempt them; but they are often wrong or, at the very least, subject to some ifs, buts, and exceptions. It takes a long time to develop a blue whale or a Californian redwood from a fertilized egg. That space of years is more than enough for a microbe with a generation time of half an hour to evolve resistance to penicillin. So the developmental and evolutionary timescales overlap, but not by much.

Both of the two great processes of biological creation have their historical heroes, though those of evolution tend to be better known than those of development. Darwin and Wallace spring more readily to most layminds than Fabricius or Roux. The heroes of genetics are important too, as genes underlie both processes; so we are also indebted to Mendel, Watson and Crick.

But I am more concerned here with the present and the future than with the past. For a quiet revolution is beginning in biology, and as yet its heroes are relatively unsung. While studies on development and on evolution were carried out in relative isolation for most of the twentieth century, today there is a thrust towards synthesis. A new interdisciplinary field is emerging, which goes by

the odd shorthand name of 'evo-devo', largely because its full title –
Evolutionary Developmental Biology – is rather cumbersome.

How can the small and rapid be productively united in a
conceptual way with the grand and slow? What do embryos tell us
about evolution, or vice versa? Is this a great leap for biology or just
another step? These are the questions I try to answer herein.

Some of the answers that I give to these and related questions
are the same as those that most other biologists would give. Here,
we are in the realm of well-established 'facts'. But some of my other
answers are new and/or controversial. In particular, the core of the
book deals with a single, controversial question of the utmost
importance: do biases in the ways in which embryos and other
developmental stages can be modified provide a sort of internal
'direction-finder' to the process of evolution that interacts with its
external equivalent, namely natural selection? This question first
arises in Chapter 2, and is explored in detail in Chapters 7, 8 and 9.
Its ramifications are considered later, especially in Chapters 16
and 17.

It is this core question – and my proposed answer of 'yes' to it –
that gives the book its name. But notice that a complication has
already crept in here in the form of 'other developmental stages'.
The idea of biased developmental variation extends beyond the
embryo, to all post-embryonic stages. So I have undersold myself in
the title; but then again *'Biased embryos, larvae, juveniles,
adolescents, etc.'* lacks a certain style.

It is more conventional to put forward new scientific ideas in
the specialist literature. And indeed I have given technical accounts
of developmental bias and the other controversial ideas discussed
herein in the appropriate biological journals. But I couldn't resist the
challenge of trying to make these important ideas accessible to a
wider audience, which is the aim of this book. I have written it in a
way that I hope will maximize my chances of achieving this aim. So
there are no Latinized species names, no mathematical models, and
only minimal genetics. The book is short; so are most of the

individual chapters. The writing style is casual. I have written the text as if it were a conversation. There is no stale scientific passive, which is increasingly out of favour anyhow. I have also thrown in a few anecdotes about my own life, where these help to introduce a particular topic in a reader-friendly way. I have avoided jargon wherever possible, but have included a Glossary to cover things that I don't think of as jargon but you might.

So I hope that the book will be accessible to biologists and non-biologists alike, though the former may wish to 'fast-forward' through some of the introductory material (in particular, developmental biologists may wish to skip Chapter 4 and evolutionary biologists Chapter 5). Equally, I hope the book will be accessible to everyone from first-year undergraduates (in any subject) through to emeritus professors. Not just accessible, but interesting too, and perhaps, in just a few places, awe-inspiring. If new ideas about the relationship between the two great processes of biological creation cannot inspire awe, what can?

Acknowledgements

Many people have generously given their time to read successive drafts, of which there have been quite a few. Generally speaking, the more reactions an author takes account of before finalizing a script, the better it becomes. In this case, I have been careful to take on board comments made by both specialist and non-specialist readers, in order to try to cross that dangerous rope-bridge between the twin chasms of inaccuracy and unreadability, without falling off in either direction.

So, I would like to thank all of the following for their noble efforts in helping me perform this balancing act. However, I should add that if my belief that I have not plunged into either chasm is perceived by later readers to be misguided, then they know full well where the ultimate responsibility for a book lies, and it is clearly not with those whom I now thank: Chris Arthur, Jack Cohen, Ward Cooper, Richard Gordon, Kenneth McNamara, Alec Panchen and Mary Scott.

Although I very much enjoy writing, especially when it 'flows', I am also a firm believer in the saying that 'a picture is worth a thousand words'. As ever, there is an issue of balance here, and if I had published *Biased Embryos and Evolution* as a 'picture book', few if any would have taken it seriously. But it would have been a drier affair altogether without the many illustrations that are scattered through it and that, I hope, make some of the ideas 'come alive'. For their design and execution I thank Raith Overhill and Tracey Oliver.

Finally, although it is less common for the authors of books to acknowledge sources of funding than it is for the authors of primary scientific papers – since book authors are rarely funded in advance by anyone other than their publishers – there is one organization that I would like to thank in this respect. In 1995, I was awarded a two-year Research Fellowship by the Leverhulme Trust. Although that was to write an earlier book (*The Origin of Animal Body Plans*, Cambridge University Press, 1997), I suspect that, had I not written that one, the present book would have been much impoverished, and indeed might never have appeared at all. This single piece of funding has had more far-reaching effects on my scientific thinking than all the other research funds that I have had over the years put together. Long may such schemes continue.

I The microscopic horse

You ask me to describe a horse; I answer as follows. A horse is a microscopic animal that is incapable of movement. It consists of a rather small number of cells (a few hundred, as opposed to the trillions found in a human). These cells are not organized into sophisticated organ systems. The horse is a parasite of another animal, and so acquires its resources from its host. It is entirely incapable of acquiring energy in any other way. There is no fossil record of its existence, so for all we know there may have been no such thing as a horse before the dawn of the art age in the caves of France, where our forebears drew remarkably good pictures of horses, among other things.

But wait. Their horses don't look like my description. And indeed since my description at first sight looks quite mad you might wish to agree with the cavemen and not with me. There is, however, method in my madness. My description is fine. It just refers to a time-slice in the horse life cycle that is different from the one we normally picture in our minds at the mention of the word 'horse'. We picture the adult, or if not this then perhaps a beautiful but unsteady newborn foal. What I have pictured is the horse as an early embryo, invisible to our view because it is implanted deep within its maternal host.

The point I am getting at here is that animals, and indeed all organisms, are four-dimensional things. The three dimensions of their bodies expand and change as they slide along that slippery and inevitable slope of time. Even as adults we change, albeit more slowly and often not in encouraging ways. As the American biologist John Tyler Bonner has put it, organisms do not *have* life cycles, rather they *are* life cycles.[1] We tend to picture adults in our minds for all sorts of reasons. Our brains handle three dimensions more easily than four. Adults are bigger and more visible. Even when developmental stages are big and conspicuous, like tadpoles, they are often short-lived compared with the adult. But this is not always so. In some insects, perhaps most famously mayflies, the adult lives a transient life of at most a few days, while the developmental stages through which it was produced lasted much longer. But even in these cases where the rationale for thinking in terms of life cycles is strongest, we still tend to picture the adult in our mind's eye.

The reason for this is rooted in language. Often, at least for familiar creatures, the very word we use may be adult-specific. A tadpole is, arguably, not a frog. But is a foal not a horse? And the same applies in the invertebrate world. A caterpillar is, arguably, not a butterfly (though its genes are identical); but a baby centipede is definitely a centipede.

Whether we should fall into the old familiar groove of picturing the adult, or whether we should be more mentally adventurous and try to force our lazy brains to go 4-D and 'think life cycles' depends on what we are trying to do. For our cave-painting ancestors, the adult was all that was required. But for understanding how horses evolve, this static picture just won't suffice. Every stage in a life cycle only comes into being if the previous one survives. An adult can only come into being if *all* the earlier stages survive. At the level of the individual, death is all too real an option at every single stage. Therefore at the level of the population there will be natural selection at every stage – because at each stage some individuals will live and some will die. Of course the living and the dying could be genetically identical and the difference merely a matter of chance. But the last century's accumulated knowledge of the huge amount of genetic variation present in nearly all natural populations suggests otherwise.

All this is beginning to sound very conventionally Darwinian. And in some ways, so it is. But Darwinism is all about mechanisms, and we are not quite ready to discuss those yet, or to consider the extent to which Darwinism is acceptable to those who take a developmental approach to evolution. First, we need to complete the mental shift that we have begun towards a four-dimensional view of organisms.

Let's consider a very simple evolutionary tree with just four species: ourselves, a cow, a hen and a fish. There are three ways we can picture this evolutionary tree, as can be seen from Figure 1. First there is the tree of adults; then there is the tree of embryos; finally there is the tree of life cycles. Which is best? The answer is the life-cycle tree. The others are three-dimensional shorthand. The embryo tree is a means to an end, not the end itself. Consideration of the embryo tree is meant to reveal the arbitrariness of using, as most folk do, the adult tree. In fact, since most species experience most mortality at young rather than old ages, it would seem more sensible for those intent on 3-D shorthand to use an early developmental stage as the basis for their tree. That is, they should use an embryo tree rather than an adult tree, because if we want to think in terms of some variants doing better than others in the survival game, this would seem a sensible place to start. Which has the higher mortality rate – tadpoles or frogs? Statistically speaking, there's no contest.

A tadpole, however, is not an embryo. Usually, we restrict the term embryo to those developmental stages that are protected from the elements by virtue of their location within their mother's body or, in some instances, within the casing of an egg. So our embryo tree is too simple. Science is all about generalizing (more on this in Chapter 6), and embryos are special cases of the more general concept of developmental stages. But then again, these stages, like the adult, have no clear boundaries. A life cycle does not operate in discrete stages – rather the process of development is a continuous one. This is even true in those cases, like the tadpole/frog, where major changes occur between one 'stage' and the next. Life flows. So, using the embryo tree as a means of forcing our thoughts out of old and inappropriate habits, we nevertheless end up not with this tree any more than the tree of adults. We end up with the life-cycle tree.

* * * *

(a)

(b)

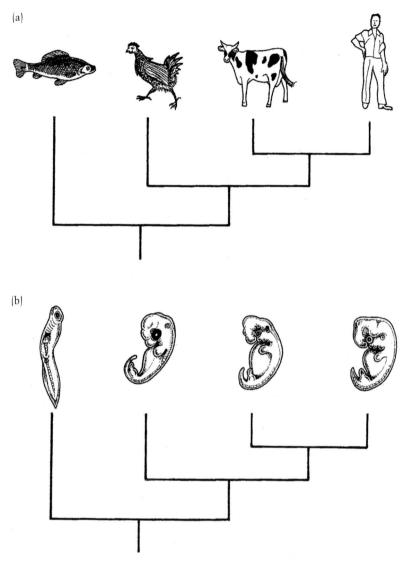

FIGURE 1 Three ways of picturing evolutionary trees: (a) adult tree;
(b) embryo tree: (c) life-cycle tree.

(c)

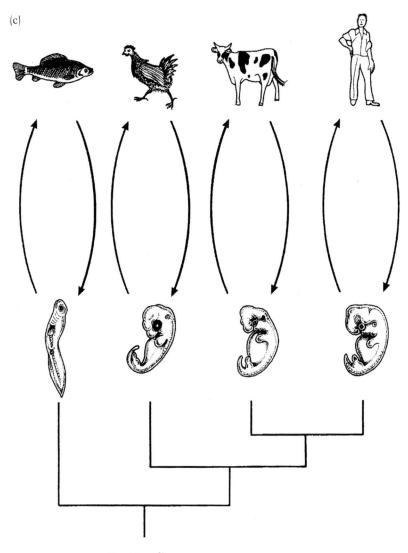

FIGURE I (Continued)

If we venture into the realm of imagination, it is possible to go forwards or backwards in evolutionary trees. And there is a major difference concerning time travel in these two directions. Consider yourself as an observer represented on any of the versions of Figure 1 by a bright red symbol (a monorail train is appropriate here) moving inexorably along the line of descent. If you move forward in time, you keep being faced with a series of decisions to make. Right fork or left? But if you start at the top of the tree, at the twig representing any of the four arbitrarily chosen extant species, and move backwards in time, no choices arise. Rather, you just keep going until you fall off the end into the primordial soup.

This picture of reverse-gear evolution is useful because it serves again to remind us that life flows continuously between generations as well as within them; over millions of years as well as periods of a few days. Begin with a human life cycle and start going back through the generations. Then accelerate. Eventually you are blasting back through ape-like life cycles, then through even less humanoid ones; and you might just notice the thin confines of a flatworm life cycle before you completely lose your head. (The first animals didn't have any.) Countless backward cycles have taken you close to the dawn of animals.

Notice that I said 'ape-like' rather than 'ape'. Present-day apes, like the chimp, are not our ancestors any more than we are theirs. No current species is the ancestor of any other. This is logically impeccable but frequently forgotten in careless discussions of evolution. Of course, following a lineage divergence, one line of descent may undergo much more profound changes than the other. But this does not mean that the comparative evolutionary slowcoach is standing still. In evolution, no one stands still for very long. Even when nothing much is happening on the outside, molecular changes are happening within.

What causes one life cycle to be different, even if just ever so slightly, from the one that went before? To consider this question, let's think of each life cycle as starting with a fertilized egg and ending with an adult. Although this disregards the chicken-and-egg problem, we need to have some landmarks to find our conceptual way, and these are as good as any. Indeed, there is one respect in which this particular mental picture is best. This concerns codes versus actual things. The fertilized egg is a minimal thing, as animals go. In our case, it is a tiny fraction of a trillionth of what will follow in terms of cell number. But from an information-content perspective it is just the same. The egg contains all thirty thousand or so genes of the recently revealed human genome. No new genes are created as we develop. Occasionally, as in mammalian red blood cells, genes are lost. But generally, the genes are simply copied each time a cell divides. So one way to look at development is as a means of getting from minimal form and maximal encoding to the opposite state of affairs through a complex interconnected series of code-readings and construction events (of which I give some examples in Chapter 4).

An offspring life cycle will thus be different from its parental life cycle only if something that affects the great developmental unfolding has changed. Such things can be of two rather different kinds: genetic or environmental. If you take a big fly that grew up with lots of great maggot-food and get it to lay an egg somewhere where the food supply is only just sufficient for growth, the new life cycle will produce a smaller fly with fewer and/or smaller cells. Alternatively, if the new food supply is the same as the old, we might still get a smaller fly if a gene involved in the production of a growth hormone has mutated. In general, it is the latter, genetic type of change that is of interest to those who study the evolutionary process. However, the two cannot always be so neatly separated. Sometimes they interact. For example, a gene mutation can alter the way a developmental process responds to an environmental change. But for now things are complex enough, so I will defer discussion of such matters until Chapter 12.

The American evolutionary biologist Leigh Van Valen once said[2] that 'evolution is the control of development by ecology'. This statement, which I believe captures only part of evolution's essence, would benefit from dissection. What Van Valen meant was that, over the years, as environmental conditions change (or as organisms invade new environments, which amounts to the same thing), those life cycles that are fittest for their environments prosper. Or, to put it another way, the environment is moulding life cycles through the agency of natural selection. Given genetic variation for life cycle features, those genes that produce what we can for now simply call fitter life cycles tend to spread, while those that produce less fit ones gradually die out.

If this is how evolution works, then to understand it we need to know about two things: the mechanics of natural selection (the Darwinian realm), and the mechanics of building bodies (the developmental realm). But if that were all you needed to know, I could simply refer you to two textbooks – one on population genetics and one on developmental biology – and you could read first one and then the other. If that were the answer, it would have saved me the job of writing this book. But while that way lies some of the truth, the whole truth is harder to acquire. In my view, a developmental approach to evolution is not simply a bolting-together exercise. Rather, it is a case of the whole being more than the sum of its parts.

What I mean by this is that juxtaposing the two great disciplines of evolutionary and developmental biology produces insights that do not emerge from either on its own – including an insight into what determines the direction in which evolution proceeds. These insights alter in a fundamental way both how we see embryos and how we see evolution. They collectively characterize the nascent field of Evolutionary Developmental Biology or 'evo-devo'. In the approach used by students of evo-devo, embryological (and larval) development becomes a front-line soldier in the battle to construct an elegant, accurate and complete evolutionary theory, rather than some straggler at the rear that everyone has long since forgotten. Let battle commence.

2   What steers evolution?

In science, as elsewhere, it is pointless to fight the same battle twice. The only way to avoid such wasteful activity is to know about the battles that have gone before. So, while I am no historian, and you may not be either, we all need to know enough about the history of evolutionary biology, and the battles that have been fought between the 1800s and today, in order to fight a *new* battle that will advance knowledge. If we fail in this task, we are merely cluttering up the literature (in my case) and our brains' limited storage capacity (both of us) with unnecessary duplication of battles that have already been fought.

Of course, I cannot do justice to the history of evolutionary biology in a single chapter. But that's no problem, really, for two reasons. First, there are lots of books 'out there' for anyone who is interested; and second, I only need to deal here with those past battles that are most relevant to the one that I have chosen to fight in the present. This concerns the forces that 'steer' or 'drive' evolution in particular directions rather than others. So the focus of my history will be just such driving forces, and how thinking about them has altered from Darwin's day to our own.

It hardly needs to be said that 'driving forces' are at the very heart of things. Evolution has been going on for three or four billion years, and has produced all the particular creatures that we see around us, including our fellow humans. It has also produced all those other particular creatures, such as dinosaurs, that are now extinct but have left fossilized remains as evidence of their once very real existence. Equally, there are many imaginable creatures, including the six-legged horse (Figure 2), which evolution has never produced, and perhaps never will, despite their apparent stamp of approval from an

FIGURE 2 The fictitious but potentially viable six-legged horse.

engineering-design point of view (five-legged horses – no; six-legged – why not?). Had other evolutionary directions been taken, a different array of creatures from those with which we are familiar would have become actualized, while others, maybe including humans, would have remained only in the realm of the possible.

If you ask an educated 'person on the street', a typical biology undergraduate, or a typical professional biologist the question 'what is the main driving force of evolution?', you are most likely to get the answer 'natural selection'. The reason for this is that generations of biologists (and lay audiences with interests in things biological) have been brought up against a philosophical background imbued with Darwinism (good) and the notion that the variation upon which Darwinian selection acts is entirely 'random' (bad). We have been educated to think that variation is not produced in response to an organism's environmental needs (good), and that the nature of variation is thus irrelevant to the prevailing direction of evolutionary change (bad), as long as there is *some* variation in the first place so that natural selection has some raw material with which to work.

In this book, I contest this overly Darwinian view of the determination of evolutionary direction. I do not contest natural selection (for more on this, see the next chapter); but I *do* contest the notion that it alone sets the evolutionary sails. In fact, Darwin himself contested this, but not in the same way that I do here. Darwin was prepared to acknowledge some role for Lamarckian processes ('use and disuse'), while I am not. Darwin did not make a detailed case for the structure of variation being an important determinant of the route that evolution takes, while I do.

My central thesis, as will become gradually clearer, is that the direction that evolution takes is determined by the interaction between two agencies: 'developmental bias', meaning the tendency of developmental systems to vary in some ways more readily than others (Figure 3), and natural selection. I do not assign a relative importance to the two, for example to say that 'developmental bias is the *main* agent of directionality' because if they function as an interacting pair then such a statement is meaningless.

∗   ∗   ∗   ∗

So now for a bit of history. And let's do it chronologically rather than in 'flashback' mode. The recurring theme to watch out for is the debate about the relative importance of external and internal factors. This issue has been referred to by Stephen Jay Gould as one of the 'three eternal metaphors' of palaeontology.[1] That is, it is a debate that has occupied the minds of palaeontologists and biologists from before the advent of evolutionary theory right up to the present day.

Here is a historical sequence of the main agencies that have been suggested, from about 1800 onwards, as driving the evolutionary process in particular directions.

First, use and disuse, or 'the inheritance of acquired characters' if you prefer. That is, the notion that an animal's striving to use some part of its body to the fullest will somehow cause that part to be enhanced, and this effect to be somehow passed on to its progeny. The giraffe's neck is the classic example, of course. Now although there have been various attempts to resurrect this Lamarckian mechanism, some of them as recent as the 1980s, none has ever stood up to experimental testing. As far as we know, there really are no mechanisms through which acquired characters can be inherited. Personally, I regard this debate as closed, and so do most other biologists. So I will

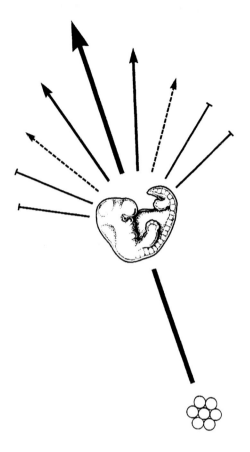

FIGURE 3 Developmental bias, represented by different probabilities of reprogramming an embryo's trajectory in different ways. Large arrow, original trajectory; solid arrow, easy; dashed arrow, difficult; blunt-ending lines, impossible.

not consider this proposed agent of evolutionary directionality any further.

Second, natural selection, as proposed by Darwin and Wallace in 1858. I don't need to say too much here, partly because the next chapter is all about selection, and partly because most folk, including non-biologists, know at least the bare bones of the story. Darwin claimed that selection was the 'main' evolutionary driver; Wallace came perilously close to suggesting that it was the *only* one. Regardless of this difference between them (of which more later), natural selection became accepted as an important (the most important?) evolutionary driver by many biologists shortly after publication of *The Origin of Species*[2] in 1859. And although there have been attempts to dislodge selection from this elevated position, none of them has ever really succeeded, at least from the perspective of influencing the whole community of biologists as opposed to a particular school.

Third, orthogenesis. This is a process proposed by several late nineteenth century biologists in which evolutionary lineages have an intrinsic tendency to go in certain morphological directions – e.g. bigger body size. The proponents of orthogenesis are not well known names, at least to most present-day biologists. Three of the leading figures during the heyday of orthogenesis (the 1870s to the 1890s) were Theodore Eimer, Carl Naegeli and Wilhelm Haacke.[3] There were many biologists during that period who did not accept orthogenesis, and in the early twentieth century it fell from favour almost entirely. The reason for this was that it lacked a demonstrable mechanism. Even the orthogeneticists themselves were not of one mind on what 'drove' lineages in particular directions. But most were internalists at heart and thought that some unspecified internal mechanism was the evolutionary driver.

A theory of evolutionary directionality based on an unspecified mechanism is clearly built on sand. Orthogeneticists were seen by many as mystics; hence their scientific demise. But, even though I have no time for mysticism, I have to admit some sympathy for their cause. My argument here (especially in Chapters 8 and 9) for the importance of developmental bias as an evolutionary driver could be seen as an argument for the internal mechanism that orthogeneticists couldn't quite put their finger on at the time. But it is a mechanism that works hand in hand with selection. There is no conflict between the two, as we will see.

The fourth proposed evolutionary driver was mutation. Ever since the rediscovery of Gregor Mendel's work in 1900, biologists have accepted the reality of mutations. But there have been major disagreements about their importance in evolution. The 'mutationists' argued that evolution normally took place in large jumps or saltations at the phenotypic level, which were underlain by big-effect mutations in the genes. This view was taken by the Englishman William Bateson[4] even before 1900; it was championed by the Dutch botanist Hugo de Vries[5] shortly after the turn of the century, and by German-American geneticist Richard Goldschmidt[6] somewhat later (the 1930s to 1950s).

Mutationism never gained general acceptance across the broad community of biologists. Instead, the prevailing view that emerged in the twentieth century – neo-Darwinism – held that evolution was a gradual process based on many tiny-effect mutations rather than a few large-effect ones. This was essentially a reaffirmation of Darwin's *'natura non facit saltum'* – one of the very few cases in which the great man seems to have departed from his usual broad-minded pluralism. Today, we can see that the truth lies somewhere in between the mutationist and neo-Darwinist views, albeit closer to the latter. Dissection of the genetic basis of variation in morphological (and hence developmental) characters usually reveals neither a single large-effect gene nor hundreds of 'polygenes' with individually negligible effects. Rather, there are usually several genes involved, the exact number varying from character to character. Most of these genes have quite small effects, but a few are often responsible for a large proportion of the variation – 'mesomutations', if you like, in contrast to 'micro' or 'macro' mutations.

Given this finding, it is important that any proposed evolutionary driver that is to have wide applicability should be capable of acting on the bigger *and* smaller changes that both contribute to the evolutionary process. Natural selection is one such driver (though of course its relative contribution to directionality is greater when it acts on smaller-effect mutations). I hope to persuade you that developmental bias is another.

Since the 1950s, three other evolutionary drivers have been proposed, but I don't think that any of them are particularly relevant to my main story here, which concerns the evolution of development. The Japanese geneticist Motoo Kimura[7] argued eloquently, through the 1960s, 1970s and 1980s, that the random process of genetic drift could often, counterintuitively, lead to directional changes. Most biologists now accept that he was right within his chosen domain – the structure of protein molecules. But Kimura himself emphasized that genetic drift would not play an equivalent role in the realm of organismic structure.

In the 1980s, when it became clear that the genome was a much more fluid thing than previously thought, with all sorts of internal dynamics of its own, including transposable elements, slippage of DNA strands and gene conversion, the English geneticist Gabriel ('Gabby') Dover proposed that evolutionary directionality could be determined by these newly discovered dynamics – collectively referred to as 'molecular drive'.[8] In my view, the status of this proposal is a bit similar to that of Kimura's; that is, molecular drive may make a contribution to genomic evolution, but most of the molecular processes included within this overall 'umbrella' have not been demonstrated as yet to have a major role in the evolution of development and the structures that the developmental process produces. Having said that, though, one particular molecular process, gene duplication, clearly does have far-reaching consequences, as developmental systems are often controlled by families of related genes that appear to have arisen through the repeated duplication of a single ancestral copy.

Running in parallel with these proposals of the importance of drift and molecular drive, which arose from genetics, were proposals of a very different sort, stemming from palaeontology. From the 1970s onwards, many palaeontologists argued (and still do) that selection acting at higher levels of organization than the individual organism is important in determining the direction of evolution. Personally, I don't find any difficulty in accepting the idea that 'species selection', as proposed by American palaeontologist Steven Stanley,[9] plays an important role in evolution. It may well explain why some clades have more species than others, for example. But I don't think that it can explain key issues in the evolution of development, such as how animals that lack limbs, hearts or brains evolve into those that possess such organs.

Finally, and most relevant to my quest, a third roughly parallel strand of evolutionary thinking began to emerge in the 1970s and 1980s, in which the developmental process itself is proposed to have an important role in determining evolutionary directionality. Prominent pioneers of this view included the Harvard-based biologists Stephen Jay Gould,[10] Richard Lewontin[10] and Pere Alberch[11]. While I believe that their contributions were seminal, I also think that they made one important mistake. They, and many others, emphasized 'developmental constraint'. Despite an attempt by Gould to indicate that he saw 'constraint' as a positive as well as a negative thing,[12] most biologists, and indeed most non-biologists too, interpret 'constraint' in a negative way. My view is that development biases evolutionary directions in both positive (drive) and negative (constraint) ways.[13] So I think that the use of 'constraint' to cover both of these is misleading. But make no mistake: although I will use a different language from Gould and Lewontin, I will be singing the same song. What follows is a continuation of their argument. I hope that many others will take up the cudgels on their behalf.

* * * *

You might have noticed that the above list of people who have proposed evolutionary driving agencies has a curious gap in it. Present-day evo-devo is in many ways a resurrection of nineteenth-century comparative embryology. It is comparative embryology in the age of the gene. Three of the most influential figures in nineteenth-century comparative embryology were Geoffroy Saint-Hilaire, von Baer and Haeckel. None of these three 'founding fathers' has yet been mentioned. There is a good reason for this: none of them proposed a credible evolutionary mechanism.

In von Baer's case, he made no such proposal because he did not believe in evolution. This is ironic in the extreme, because his famous observation that the embryos of different types of vertebrates get progressively more different as they develop,[14] published in 1828, cries out for an evolutionary explanation. Yet even after publication of *The Origin of Species* in 1859, von Baer refused to accept either natural selection in particular or evolution in general.

Paradoxically, Geoffroy Saint-Hilaire[15] did become an evolutionist, despite the fact that he died before *The Origin of Species* was published. And he did propose an evolutionary mechanism, but it was a rather fanciful one based on a putative decline in the amount of 'energy' in atmospheric oxygen. So he is not remembered for what he said about mechanism, but rather for what he said about pattern. Most famously, perhaps, he proposed (in 1822) that vertebrates were a sort of 'upside-down' version of some invertebrates, because (among other things) they have a dorsal nerve cord and a ventral heart instead of the alternative arrangement that can be found, for example, in an earthworm or a fly. This looked fanciful to many biologists too, both in his own century and the next. But in the 1990s comparative developmental genetic studies revealed dorsalizing genes in vertebrates that were ventralizing in insects, and vice versa. So it looks as if Geoffroy was right after all, but it took almost 200 years for the relevant genetic evidence to arrive.

Haeckel, working somewhat later than the other two, and publishing his most famous work[16] post-*Origin* (in 1866), was both an evolutionist and a Darwinian. He believed in natural selection. Indeed he believed, as do biologists today, that natural selection works by modifying the course of development. And he set about describing the patterns that he thought this modification would produce. His main efforts went into attempting to establish the prevalence of the pattern known as recapitulation: that is, embryos of 'higher' animals going through stages that resemble the forms of their 'lower' animal ancestors, and thus, in a sense, carrying an embryological record of their evolutionary ancestry.

Haeckel's recapitulation has often been misrepresented, and he himself unfairly maligned, by many subsequent authors.[17] He is portrayed as if he thought that a human embryo, for example, went through developmental stages that resembled the *adult* forms of 'lower' vertebrates. Yet he speaks of human embryos as resembling the 'undeveloped embryo form' of other species.[18] He is portrayed as anti von Baer, yet he sings von Baer's praises. The fact that von Baer saw 'divergence' where Haeckel saw 'recapitulation' does not represent a contradiction. It is just that von Baer, being a non-evolutionist, was never thinking about ancestors; rather his comparisons were between present-day animals. But Haeckel, as an ardent evolutionist in the era of evolutionary excitement that followed publication of *The Origin of Species*, was often thinking about ancestors. Divergence of two present-day taxa and recapitulation of each in different directions from a common ancestor are perfectly compatible with each other.

Of course, Haeckel was quite wrong to try to turn a pattern into a 'law'. Statistically, evolution seems to modify later developmental stages more often than it modifies early ones. And, where it adds bits to the developmental programme, it may, again statistically, add these more often near the end than near the beginning for perfectly sensible selective reasons. But statistics is just that. It deals with probabilities, not the certainties that are normally associated with scientific 'laws' (such as $E = mc^2$). Evolution goes in many directions. Sometimes a lineage gradually moves towards greater organismic complexity. In such cases, we would expect to see elements of recapitulation, albeit imperfect and accompanied by other types of change. But when a lineage goes in the opposite direction, as is often the case with parasites, recapitulation would hardly be expected, as development is more likely to experience subtractions than additions.

* * * *

So, we conclude that nineteenth-century comparative embryologists were better at revealing interesting patterns than in coming up with evolutionary orienting agents or 'drivers'. Given the circumstances of the times, and the kind of data at their disposal, this hardly seems surprising. But what of their later counterparts? Although there was something of a gap between the demise of old-style comparative embryology around 1900 and the rise of new-style evo-devo in the

1980s, this was a gap in relative terms. Individuals working on, effectively, the evolution of development were few and far between rather than non-existent. Let's look at whether any of them had anything to say about evolutionary mechanisms.

The great D'Arcy Thompson, who produced his magnum opus *On Growth and Form*[19] in 1917, did not. He, like his predecessors, was more interested in patterns than in mechanisms. He took a few side-swipes at the idea of natural selection acting on single characters, because he saw organisms as integrated wholes rather than collections of independent characters (ironically, Darwin would have agreed). However, he refrained from proposing any driving force of his own for the evolutionary process. Nor are figures such as Julian Huxley[20] or Gavin de Beer[21] of particular note in this respect, because they simply took natural selection on board and used it to try to explain certain kinds of developmental pattern. It was not until the 1940s and beyond that a small number of biologists, most notably Schmalhausen[22] and Waddington,[23] began to think about development as something other than a passive player in the evolutionary game, a sort of putty at the receiving end of natural selection, with no influence on evolutionary directionality.

Waddington's concept of genetic assimilation is especially worthy of mention in this context. He showed that under certain environmental conditions development would alter its course in particular directions, and that by selectively breeding from those organisms most prone to this 'deflection' of developmental trajectory, it was possible to get, eventually, to a point where it would occur spontaneously. Although at first glance this might look like a Lamarckian process, it is nothing of the sort. Rather, it is a process in which the nature of developmental variation, and its interaction with the environment, contribute to the direction of evolutionary change *in conjunction with* selection. In my view, the proposal of (and experimental demonstration of) genetic assimilation was a milestone in the history of evolutionary biology, and its importance deserves to be more widely recognized than it is.

Waddington published important papers and books through several decades, ending in the 1970s with his collection of essays entitled *The Evolution of an Evolutionist*.[24] The end of the Waddingtonian era was followed by what I choose to call the 'modern era', from 1980 to the present.

* * * *

So, let's now broaden out and look at the overall structure of evolutionary biology, as it exists today. This is especially important in one particular respect, relating to what is often called 'mainstream' evolutionary theory. Does such a mainstream exist? If so, am I working within it or arguing against it? If not, what sort of disarray is evolutionary theory in, and how do I deal with such a situation?

It's probably true to say that there is not even agreement across the whole community of biologists on the correct answer to the first of these questions. Some see neo-Darwinism as the mainstream theory; others reject it completely and see something else as mainstream; others argue that neo-Darwinism was the mainstream for half a century or so, but that the river has now become a delta and there are lots of parallel streams, of which neo-Darwinism is just one. Others declare themselves agnostic. Given this untidy situation, what I will do is to give you my personal view of the structure of present-day evolutionary theory.

My own view corresponds to the third option given above – the delta. That is, I believe that there is, to a certain extent, a state of disarray, with various schools of thought each going about their business and largely ignoring the others. And indeed each 'school of thought', inasmuch as these can be delimited, is itself heterogeneous. I will describe three main schools below, and will attempt to identify for each both its defining features and the variation within it.

* * * *

First, the 'phylogenetics' school. Here, the emphasis is on attempting to find the true tree of life (or at least of some corner of the living world). That is, the thrust is towards determining which routes evolution took in the past, and which creatures are most closely related

WHAT STEERS EVOLUTION? 21

to which other creatures. So the emphasis is on pattern rather than mechanism. The latter takes a back seat (at best) in this endeavour.

But those who focus on phylogeny come from many different backgrounds. Some are essentially theorists and are most interested in refining the methods of constructing and testing trees. In this group, there are a few extremists who are almost anti-mechanism, taking a strong 'pattern before process' view. Their argument is that there is little point in trying to establish how evolution is 'driven' until we know which route it has taken. This raises the question of the extent to which the two things – pattern and mechanism – are independent. A reasonable counter to the 'pattern first' view is that some mechanisms, natural selection being a prime example, transcend any particular pattern of relationship, and can be studied in parallel with studies of evolutionary relationships, rather than only after the latter have been completely clarified (which in itself is perhaps a rather optimistic goal).

As ever, extremists are in the minority, and there are far more 'theoretical phylogeneticists' who, rather than being anti-mechanism, simply regard studies of evolutionary mechanisms as being in someone else's domain. This seems to me entirely reasonable. Both endeavours – studies of pattern and of mechanism – are sufficiently daunting that specializing in just one or the other may be a sensible approach.

The second group of phylogeneticists is composed of all those who study the morphology of various creatures and use that information to try to find the true tree of relatedness for the creatures that they study. Of course, there is overlap between this group and the last; and indeed overlaps abound everywhere among the various schools and subschools within evolutionary biology. (Please take that as read from here on, as there is little point in my repeating this each time I move on to the next school.) The morphology group comprises both palaeontologists and those who study the comparative morphology of extant creatures. Most members of this group seem to quietly acknowledge the reality of natural selection as an evolutionary driver, but, like the last group, they don't place much emphasis on it as their primary interest lies in the domain of pattern rather than process. However, it is worth recalling that some new ideas about mechanism – such as 'species selection' – have arisen from recent palaeontological studies (from the 1970s onwards), and also that many palaeontologists have made much of 'heterochrony' (of which more later).

The third group of phylogeneticists is of much more recent origin. I am referring, here, to those who study comparative molecular biology. The information with which members of this group work – DNA and protein sequences – is very different from morphological information. Nevertheless, the use to which it is put is the same – attempting to discover the true pattern of relationship within a particular group of creatures. This includes both high-level groups (like the animal kingdom) and lower-level groups (such as the mammals). In general, the more slowly that a particular gene evolves, the higher the level of taxonomic group that it can be profitably used to tackle. Molecular phylogeneticists, like their morphological counterparts, are usually quiet supporters of natural selection, being primarily students of pattern rather than mechanism.

*　*　*　*

We now move back up a level, from subschools to schools. The second school of present-day evolutionary thought that I want to briefly outline is the 'population' school. Here, the focus is very much on mechanism rather than pattern. And most members of this school are firm believers in natural selection as the main or sole driver of adaptive evolution. I use 'adaptive' here because the reality of genetic drift is now widely accepted too, but the component of evolutionary change that it 'drives' is non-adaptive.

The population school, like the phylogenetic one, is heterogeneous. This time, I want to distinguish two main subschools: population genetics and behavioural ecology. Population genetics is, in some usages, almost synonymous with neo-Darwinism. Population geneticists study natural selection in detail from theoretical, experimental and observational perspectives. Population genetics *sensu stricto* tends to be reserved for the theoreticians who work with mathematical models. This group has its origins in the work of the early twentieth century 'triumvirate' of the Englishmen R. A. Fisher[25] and J. B. S. Haldane[26] and the American Sewall Wright.[27] But population genetics *sensu lato* also includes other, related types of study. Observational work carried out in the wild, such as the many studies of colour patterns in butterflies, moths and snails of the 1960s and 1970s, tends to be referred to as ecological genetics. And artificial selection work using continuously variable characters such as body length or weight, often in model systems such as fruitflies or mice, tends to be labelled quantitative genetics (of which more later). In all cases, selection is central to the endeavour.

Since the 1980s, there has been a discernibly separate population subschool known as behavioural ecology; and, associated with this, sociobiology. This school, as its name suggests, is delineated primarily by the type of characters that its members study: behavioural rather than morphological or molecular ones. Also, although the behaviour of an animal, like its structure, is underlain by genes (as well as environmental influences), most behavioural ecologists operate primarily at the phenotypic level and devote little effort to genetic dissection of the behavioural traits that they study. A typical investigation by a behavioural ecologist might be a study of the breeding behaviour of a species of bird, including, for example, an attempt to quantify the degree of mate fidelity and the extent to which this variable (from monogamy to promiscuity) is related to fitness.

Early studies of this kind were dangerously close to being pan-selectionist, and even pan-externalist. That is, everything was seen in terms of selection; and not just that, but selection that was directly related to external environmental factors such as food, predators and mates. More recently, however, many behavioural ecologists have begun to temper their selectionism with thoughts of constraint that broadly parallel the ways in which morphologists think of such things. And indeed, for all their differences, behavioural and morphological characters probably evolve in broadly similar ways, under the control of the same range of 'drivers'.

* * * *

And so, finally, to the evolutionary developmental biology or 'evo-devo' school – the one in which I now mainly operate, despite an academic upbringing in the population school. The largest subschool here is that of comparative developmental genetics, where the main focus of attention is the study of a particular gene, or a group of genes, in several different creatures – often belonging to different higher taxa such as different classes or even phyla. Studies of this kind (discussed in Chapter 15) have revealed a wealth of fascinating new and unexpected facts. It has become clear, for example, that homologous morphological characters sometimes have different genetic and/or developmental bases in different creatures. Equally, homologous genes in different creatures may do entirely different jobs. This is not always the case, of course, but the fact that it *can* happen has shattered any nice neat illusions that some of us may have had, for example that homologies at the different levels would have simple interrelationships.

Other subschools of evo-devo are beginning to emerge. There have been some amazing recent discoveries of fossil embryos from the distant geological past. So there is now an embryological, as well as a post-embryological, dimension to what might be called palaeontological evo-devo. It is still true that adult fossils vastly outnumber embryonic ones in the fossil record of most groups; nevertheless the availability of some embryonic and post-embryonic developmental stages of extinct creatures is enormously valuable as a window, however small, into how these creatures developed. This helps us to see them, like their extant counterparts, as four- rather than three-dimensional entities.

A theoretical subschool of evo-devo is also beginning to emerge. This is a hugely exciting endeavour, which I suspect will grow rapidly over the next decade or so. Since developmental genes do not act alone, but rather in interacting pathways or cascades, there is a whole dynamics 'out there' (or perhaps more accurately 'in there') awaiting study.

To date the number of studies of this kind is tiny. But the challenge posed by the quantitative aspect of how these genes interact, and how these interactions and cascades evolve, is truly huge. If there were crystal balls that we could look into to glimpse the future of science, this is one area whose image I suspect we would see shining brightly.

* * * *

With regard to evolutionary mechanisms, most students of evo-devo, whichever of the three subschools they belong to, have little if any difficulty in accepting natural selection as an evolutionary driver. And some are perhaps content with that. But many of us feel that something is missing; that selection is not enough; that the actualization of some creatures, together with the failure of others to emerge from the realm of the possible, requires something else – something internal that interacts with selection in a particular way. That is what Gould and Lewontin were saying more than twenty years ago. And it is what I am saying in this book. So now we begin our quest for a more complete grasp of what orients evolution than has been available in the past. And, as with all sensible quests, we start on reasonably well-trodden ground before setting off into the unknown. So the next chapter focuses on Charles Darwin and the concept of natural selection. From there, our journey will meander in various directions but will ultimately lead to a synthesis of evolutionary thinking that is significantly broader than the 'modern' synthesis of the twentieth century.

3   Darwin: pluralism with a single core

In the course of his lifetime, from 1809 to 1882, Charles Darwin wrote several very different books. Some of them, like his monographs on barnacles, were in the 'worthy but dull' category. But one of them – *The Origin of Species*[1] – changed our view of the world. I have read it from cover to cover twice, my two readings being separated in time by about a decade. During this decade – from the early 1970s to the early 1980s – my scientific interests had undergone a major change, from the interface between evolution and ecology to the interface between evolution and development. Because of this change, the two readings were more like reading two different books. The things I noticed second time round had been invisible on my first run through, while the things I had noticed first time round had receded from view by the time I felt compelled to read this extraordinary book again, and were barely noticed on that later occasion.

This 'two readings becomes two different books' syndrome is a manifestation of an ancient truth that has been memorably put, though in very different ways, by French microbiologist Louis Pasteur and American folk singers Paul Simon and Art Garfunkel. Pasteur once famously commented[2] that 'chance favours only the prepared mind'. That is, although luck is a rather random thing, a piece of potentially important information that comes along fortuitously is likely to be overlooked by everyone except those who are in some sense, because of their general interest or their previous studies, pre-disposed to recognizing its importance. In *The Boxer*, Simon and Garfunkel sang their belief that 'a man hears what he wants to hear and disregards the rest'. And for all women's supposed (and perhaps real) advantage when it comes to 'multitasking', I suspect that they are no more immune to this selectivity of hearing (or seeing) than men.

* * * *

So, back to *The Origin of Species*, and my two selective readings of it. How did my take-home messages differ on the two occasions, and what can be learned from this difference? Well, it will probably help if I put each reading in the context of the stage of my life that I was at when each took place.

I first read Darwin's great masterpiece in the early 1970s, while on a 'year out' between gaining my first degree from the University of Ulster, based in a small northern Irish town called Coleraine, and going on to study for a doctorate at the University of Nottingham, in the English Midlands. During this year of experiencing the so-called real world, I did a variety of things. At the time I first read *The Origin of Species* I was working as an Assistant Guillotine Operator in Belfast. Happily, this position was not quite as it might seem from the job description. Despite all of Ulster's political problems, no one on either side was advocating that they could be solved by using ancient French execution techniques on leading members of the other tribe. Rather, my guillotine was located in a factory, and its sharp blade was designed for cutting pieces of steel, not human necks. So it was not a macabre job, but it was certainly a very boring one. A little reading in the evenings was necessary to keep my brain from atrophying, and what better for a biology student than Darwin's great book.

The Ph.D. project that I was about to embark on in Nottingham was in ecological genetics. As I mentioned in the previous chapter, this is the study of natural selection in action in populations of living creatures in the wild. It was born from the realization that, while evolution normally works very slowly, and so is best studied by looking at series of fossils extending through vast periods of time, it sometimes, for various reasons, works much more quickly, and so can be studied as it happens here and now. Often, this rapid evolution occurs in response to extreme environments, where 'evolve or die' are the only

two options open to the creatures concerned; and often, such environments are created by human activities. The evolution of resistance to
insecticides by the populations of insects exposed to them is a classic example, though not the one that I was going to work on, which
involved snails. But since the evolutionary processes in question are
general ones, the exact species being studied is immaterial.

Because my anticipated research project was to be based on natural populations, the messages I took out of my first reading of *The
Origin of Species* were all in this area. So what made most impact
on me was what Darwin had to say about: the variation that is found
within species in nature; the 'struggle for existence' that all creatures
experience; the way that this struggle moulds the available variation,
in other words 'natural selection' – the core of Darwin's world view;
and the way in which species tend to diverge in character, despite common ancestral features, because populations living in different places
where ecological conditions are different tend to evolve in different
ways.

*   *   *   *

Let's jump over the next ten years or so of my life and look at the context for my second reading of *The Origin of Species* in the early 1980s.
I had now, finally, obtained a permanent academic position after the
inevitable period of leaping around the country (the world, in some
folk's cases) on temporary contracts that is the usual start of an academic career. I had ended up in the north-east of England, which is an
interesting mixture of urban sprawl and beautiful, unspoilt hilly countryside. Because my home was on the edge of Newcastle and my job
in Sunderland, the twelve-mile train journey that separated the two
went through the sprawl, not through the hills. This apparently unfortunate fact was a blessing in disguise. Personally, I enjoy idly looking
out of train windows at wild hilly country; but I have no interest at
all in repeatedly staring, day after day, at warehouses, railway sidings,
or endless anonymous rows of terraced houses that, though home to
the people who live there, are for me as a transient observer just a blur
of brick and slate.

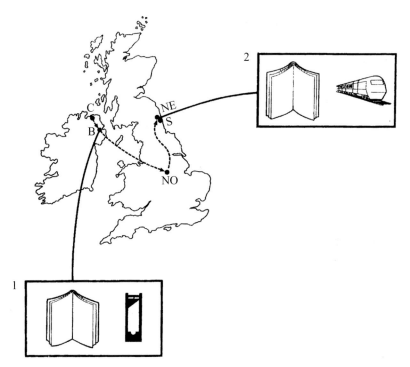

FIGURE 4 Locations of my first and second readings of Darwin's *Origin of Species*. For explanation of guillotine and train, see text. C, Coleraine; B, Belfast; NO, Nottingham; NE, Newcastle; S, Sunderland.

So what better to do than to have another read of that wonderful book? It probably took me about a month. It's a short train journey (Figure 4) and I'm a slow reader. But before I tell you about the very different messages that to my surprise emerged second time round, I need to tell you about the change in my scientific interests that had taken place over that missing decade.

I'd like to tell you a story of a gradual shift in my interests based on careful study of many learned works, my deep analysis of which took me in a definite, albeit sideways, direction. But nothing could be further from the truth. It was all a strange accident. I was looking, in Newcastle University library, for a paper[3] in *Scientific American* by the famous Japanese geneticist Motoo Kimura. This paper was in the field of population genetics, and had I found it, it would merely have given me a (good) popular account of Kimura's work, with which I was already familiar from reading his more technical articles. But strangely, I never found it. I say 'strangely' because I had gone to the library equipped with what I thought were all the necessary details. I knew the year in which it was published (1979), I knew which volume of the journal it was in, and I even knew the page numbers. So I got the appropriate volume off the shelf and turned to the appropriate page. I found myself staring at a large colour picture of a fly, which was part of an article[4] written by three developmental biologists – Spaniards Antonio Garcia-Bellido and Gines Morata, and Briton Peter Lawrence. Wrong authors, wrong nationalities, wrong paper.

The reason for this odd outcome was that *Scientific American*, unlike most journals, does not number its pages sequentially through all the issues of the year concerned; rather, it starts back at page 1 for each separate monthly issue. So when twelve of these issues got put together to make a hard-bound volume for 1979, there were twelve page 1s rather than just one of them. More importantly for me, as it happened, there were also twelve page 94s. I found the wrong one of these, which turned out to be a fortunate mistake. Of course, I like to think that Pasteur's view of chance favouring the prepared mind played a role here, but it's hard to be sure now, more than twenty years later, whether that was really true.

The paper on how flies develop was so beautiful, both in its art-work and in its science, that I was immediately captivated by it and read it straight through. It was all about the role of genes in control-ling how development proceeds, and about how flies are prefabricated in the sense of being made in a series of quasi-autonomous com-partments. Although it was a paper about development rather than evolution, I could not help connecting it with my own evolutionary interests. I was struck by the fact that evolutionary theory seemed not to include much about how the characters that varied in the wild and were subject to natural selection, such as birds' beaks or the colours of snail shells, came into existence during the development of each individual. Yet surely that must be of the utmost importance. If a character does not emerge until late in a life cycle, it will be invisible to natural selection for all those early stages during which much mortality occurs. Worse still, if the genes that control the development of the character perform other developmental tasks too, then selection for bigger beaks (for example) may be impeded because it also causes the production of smaller, less all-seeing, eyes. This is perhaps a rather unlikely combination, but it serves well to make the point.

These are just two of the many questions that sprang to my mind, on reading this 'accidental article', about the relationship between development and evolution. I had become, in a single day, a student of a field that did not yet exist – evo-devo. But did evolutionary theory from the 1850s to the 1970s really have nothing to say about development, or had I just missed it? Had my population genetics upbringing caused me to concentrate on the flow of genes in populations rather than the flow of changing organismic form as the development of an individual proceeds, and the evolutionary relevance of this second flow? The answer, as I now know, is a mixture of the two. Evolutionary theory up to the 1970s had indeed included more on development than my selective reading of it had suggested, but most evolutionists had included far too little on development in their work, and had unwisely ignored a major component of the more complete theory of evolution that is only now beginning to emerge.

So, you now know enough about my mental state at the time of my second reading of *The Origin of Species* to understand why I was predisposed to see different things; to take from it different messages; to make my reading on the train an altogether different experience from my reading at the end of long hard days working at the guillotine some ten years earlier.

The main points that leapt out at me this time were these: 'correlation of growth', which was the phrase that Darwin used to refer to the way in which different characters tend to co-vary because of a shared developmental origin; the importance of the great groups of animals (his 'classes', our phyla) with obvious similarities of body plan among the species within a 'class' but not between 'classes'; the

strong similarities that are often seen between the embryos of animals whose adult forms are very different; and the fact that, within
a species, some variant forms make their presence known at an early
stage in the life cycle, others much later. What a shift of emphasis! All
the messages of natural history that had seemed so prominent before
went by in a blur, dimly acknowledged. In their place, what shone
out like beacons were messages about embryos, life cycles and body
plans.

*   *   *   *

We have now reached a point where we can directly confront the
apparent contradiction in the chapter title. Darwin was a pluralist for
two rather different reasons. First, as can be seen from the contrasting
emergent messages from my two readings of *The Origin of Species*,
Darwin was very inclusive in his coverage. In fact, my story of his
book so far understates this inclusivity. As well as his ecological and
developmental chapters, there are chapters on geology, geography and
behaviour. *The Origin of Species* is as impressive for its extensiveness
as for its intensiveness (and, for that matter, its readability). Second,
when it came to the mechanism of evolution, where his real focus of
interest lay, Darwin elaborated his main concept – natural selection –
in great detail, and left us in no doubt of the huge importance he
attached to it as an evolutionary 'driver', yet he did not attempt to
persuade us that it was the only such driver. He famously said, at the
end of his Introduction: 'I am convinced that Natural Selection has
been the main but not exclusive means of modification.'

But Darwin the pluralist coexisted with Darwin the man with
the big idea. There is one important feature of the layout of *The Origin
of Species* that I have not yet emphasized. The messages I extracted
in my two readings do not receive equal coverage. Because natural
selection was the core of Darwin's world view, the first four chapters
are basically one long story leading up to, and including, his explanation of natural selection. This initial, substantial and impressive
block of text is the heart of the book. So the ecological messages that
I extracted on my first reading were visible to me not just because

of my own personal interests at the time, but also because of the prominence Darwin gave them. The developmental messages that I picked up ten years later were lurking, half-hidden, in small sections of Chapters 5 and 13. They were there all right, but they were not there in the same full glory as the thorough, up-front treatment of natural selection.

I don't want to criticize Darwin too much. It is obvious that anyone writing a book about a new and exciting idea will give more prominence to that idea than to other things. Also, it is true that in the 1850s much more was known about natural history than about embryology, especially its developmental genetics component, which was yet to be born. And Darwin's own personal experience, for example on his round-the-world trip on HMS *Beagle*, was largely of natural history, including the variation that he observed within species, primarily among adults; he was not an experienced embryologist. So the fact that his core idea of natural selection shone out more brightly than his relatively brief comments on embryology is no surprise, and should not be held against him. But it did have a rather unfortunate consequence in the twentieth century.

* * * *

The history of evolutionary biology in the hundred years that followed publication of *The Origin of Species* was both eventful and bizarre in its twists and turns, especially in the period just after 1900, when the seminal work on genetics by Austrian monk Gregor Mendel, published but largely ignored in 1866, was 'rediscovered'. This led to a flurry of activity in which some biologists actually thought that Mendel's work, rather than cementing Darwin's theory, could be used against it in an alternative theory of evolution by big jumps rather than small steps. But I've already discussed this in the previous chapter so now I'm going to move ahead to the period from the 1930s to the 1960s, when one thing happened (albeit in stages) that made such an impact, both good and bad, on subsequent evolutionary thinking that even non-historians cannot ignore it.

This was the formation, from the foundations that Darwin had laid down many years previously, of the school of thought that is known as both neo-Darwinism and the 'modern synthesis'. (Although most authors, myself included, use these two terms interchangeably, some do not.) This synthesis had both theoretical and practical strands, and they appeared in that order. In the 1930s, the mathematical theory of population genetics was formulated by R. A. Fisher, J. B. S. Haldane, and Sewall Wright. This theory was all about how genes would spread in populations, and thus how evolution would come about at that level. It was, if you like, Darwinism for mathematicians. This was followed, in the 1940s, 1950s and 1960s, by a plethora of studies on natural populations that were attempting to validate the mathematical theory of Fisher, Haldane and Wright. Influential figures here were the Russian émigré to the USA Theodosius Dobzhansky,[5] his German-American colleague Ernst Mayr,[6] and the English geneticist E. B. Ford.[7] The American palaeontologist G. G. Simpson[8] added a fossil dimension to this work.

In my view, the modern synthesis was both a triumph and a curse. It was a triumph because it succeeded in demonstrating how populations evolved to adapt to their local environments, and how this 'microevolution' could provide a basis, when compounded over long enough periods of time, for the appearance of new species. It was a curse because many pro-synthesis biologists attempted to portray this as the whole of evolution. There was (and still is in some cases) an arrogance about the synthesis that was entirely absent from Darwin's beautiful book. His pluralism had been lost; natural selection was king. As an undergraduate, I bought a book that actually defined evolution as a change in the gene frequency of a population. Darwin would have been horrified.

* * * *

One way of stating my central aim herein is this: it is to probe the developmental component that should have been, but was not, a major feature of the modern synthesis. It is to try to establish the form this component should take, and the degree to which its incorporation will change what went before. I ask: does this new developmental contribution to evolutionary theory come in the form of a bolt-on extra that produces no stresses in the theoretical edifice to which it is being added? Or, alternatively, does it produce a sea change in the way we view the overall process of evolution, and thus the way we understand how the whole of life on planet Earth has come about? I will be arguing for the latter view, but not in such a way as to downplay the importance of natural selection.

A word of caution is perhaps a good idea at this point. Scientific schools of thought are not frozen monolithic entities. Rather, they exhibit considerable heterogeneity both in space and in time. So when I criticize the modern synthesis, I am criticizing the work of many people in many countries over several decades. The criticisms are more appropriate for some versions of the synthesis than others. For every deficiency I point out, there will be many neo-Darwinians whose personal versions do not deserve my criticism.

But the other side of this coin is that any influential school of thought carries with it a predominant ethos. It conveys a particular way of looking at the world that emphasizes some things at the expense of others. If this 'expense' becomes too great, if the central ethos constrains too much further progress in human understanding of the process concerned – in this case evolution – then the school of thought must be criticized in the appropriate way notwithstanding the fact that not all of its adherents subscribe to all of its faults. So, with my apologies to all those enlightened neo-Darwinians out there who are undeserving of what follows, here goes.

The modern synthesis has restricted our horizons in two important ways. The first is that it has emphasized destructive forces at the expense of creative ones. Natural selection, which in the end is a force that works in terms of differential destruction, is king; mutation, in whose molecular intricacies the first moments of the creation of biological novelty are to be found, is acknowledged to be important, but then largely ignored. Development, which is how any mutation ends up affecting the organism rather than merely its genome, is paid even scantier attention. Until recently (see Chapter 7) there was not even any accepted cover term for all the ways in which development changes during evolution, in the way that 'mutation' is a cover term for how genes change, and 'natural selection' a cover term for how populations change.

How can a theory of evolution that purports to explain how creatures with trillions of cells arose from unicellular beginnings lost in the mists of pre-Cambrian time be taken seriously if all it tells us is that differential rates of destruction can alter the genetic composition of populations? How are the new variants that natural selection spreads through populations created in the first place? Although the phrase 'creation science' carries disreputable connotations because of its frequent use by some religious fundamentalists, we truly need some 'creation science' (in the other sense of that phrase) as a major component of evolutionary theory.

The second way in which the modern synthesis has limited our horizons is that it has paid too much attention to the interactions between organisms and their environment, and too little attention to the many and varied interactions between body parts that occur within each organism. To put it another way, the synthesis has been too externalist, even pan-externalist, in the hands of some of its supporters. In its more extreme forms it represents a triumph of just-so stories about adaptation to the environment over attempts to include not just organism–environment adaptation but also within-organism coadaptation. Textbooks of evolutionary biology have for years trotted out the usual old stories about how birds' beaks evolve to match their food items, or how moths' colours evolve to match their background. But where are the equally detailed studies about the importance of one body part matching another? If you peruse the literature for long enough you will surely find some lurking in quiet corners here and there. So it is not as if such studies are completely absent. Rather, it comes back to the impact that they have had on the prevailing ethos of the synthesis. And this, regrettably, has been very slight.

*   *   *   *

So, the synthesis focused too much on destruction and the external environment and not enough on creation and internal coadaptation. But this does not mean that we should seek to replace it with a body of evolutionary theory that has the complementary biases – as some biologists have tried to do in the past. That would hardly make sense. Rather, we should seek to expand the synthesis so that it becomes a more complete evolutionary theory with greater explanatory power.

Well, at first glance this all looks like sweetness and light. I am going to try to help the supporters of the synthesis by making their theory even better. They would hardly have any problem with that, except perhaps with my apparent arrogance in thinking that I can achieve further improvements after so many have gone before. This, you might say, is hardly controversial stuff.

You would, however, be wrong. Here is the crunch. There are two aspects to the incorporation of a developmental component into evolutionary theory. From the perspective of an evolutionary biologist, one is controversial, the other not. Herein, I am going to cover both. The uncontroversial bit is the input of information on how characters that we observe evolving in a population or species develop through embryogenesis (and post-embryonic growth too) in the individual organisms of which the population or species is composed. No one would object if I tried to improve the Darwin's finches story by conducting appropriate research into the genes that control beak shape and the way in which they produce their effects. That would simply make a good story even better by inserting an interesting missing chapter.

The other, controversial, aspect of the injection of a developmental component into evolutionary theory relates to what determines the direction that evolution takes. This is, after all, the heart of the matter, because the direction that evolution takes determines which life forms come into existence and which do not. The directions that evolution took in the past led one of many lineages to become human. Had different directions prevailed, I might not be here to write this book nor you to read it. The directions that evolution takes in the future will determine who or what replaces us as the intelligentsia of the biosphere.

The usual neo-Darwinian view on this subject is simple: the direction of evolution is determined by natural selection. Darwin's pluralism has been lost. He was prepared to be cautious and accept that agencies other than natural selection may have a directional role. Of course, present-day neo-Darwinians could be said to remain at least slightly pluralist because most acknowledge that variations that do not affect the survival or reproduction probabilities of the organism will increase or decrease in their relative frequency through the random-walk process of genetic drift. But this is hardly much of a concession to those who doubt the unqualified supremacy of natural selection, because by definition genetic drift only affects the direction of evolution of things that don't matter to the organism in the first place.

The view that I take here is that mutation and development are both important determinants, alongside natural selection, of the directions that evolution takes. All three are important, but I am not going to assign a ranking of their *relative* importance. In any event, I believe that it is the interactions between them that are crucial, rather than one of them acting in isolation.

It is hard to overstate how much this point of view differs from that of a pan-selectionist, as the most hardened neo-Darwinians have been labelled. Mutations of genes, and the changes in development that they cause, are contributors to evolutionary direction, not merely passive agents that serve up a sort of mush that natural selection on its own moulds into shape. For now, this is just an assertion. Arguments in favour of its acceptance will follow in Chapters 8 and 9, and will involve the mental leap from thinking only about the actual world of creatures that we see around us to thinking also of other possible worlds with other creatures that might have been. Only against this background do the deficiencies of a pan-selectionist approach to evolutionary directionality become starkly evident. Only then can we glimpse the creative roles of mutation and development that have long been downplayed or denied by many proponents of what has until recently been seen as 'mainstream' evolutionary theory. A school of thought that is characterized by this downplaying or denial deserves to be mainstream no longer. I could argue that it never did. Well, that's an academic point, as they say, because we can't change the past. But the future is another matter entirely. So let's make our way into it on the ever-shifting interface of the present, and see what happens.

4 How to build a body

Ease of study has been one of the main criteria that have guided developmental biologists in their choice of creatures. If we are interested in understanding the general principles of how organisms develop, we can in theory choose any creature we please. There is, however, the caveat that we may end up unwittingly choosing one that has peculiarities of development that render attempts to use it as a basis for generalizations misleading or just downright wrong. But this is a risk worth taking. Research by its very nature is always probing in the dark, and the risk of probing in what later turns out to be the wrong direction is always with us. If we don't take this risk, we never venture out into any sector of the darkness of our ignorance, and we never learn anything.

So, most research in developmental biology is based on what have become known as model systems, in other words a small select bunch of species that are particularly amenable to study, and that can, with luck, be used as 'models' of how development works generally. The main model systems that have been, and are being, used in this way are the mouse, the chick, the frog, the zebrafish, the fruitfly, the roundworm and, on the plant side of the great divide, a little weed called thale cress (Figure 5). We know much more about the development of these seven creatures than any others. And of course this model system approach is self-reinforcing. If you dig a hole with the aim of reaching down to bedrock, take a coffee break after the first hour, and then start digging again, what would you do – continue with the same hole or start another? It's the same with science. Once you have begun to find out about some system, it's much easier to stick with it and build on your accumulating stack of knowledge than to switch systems and start anew.

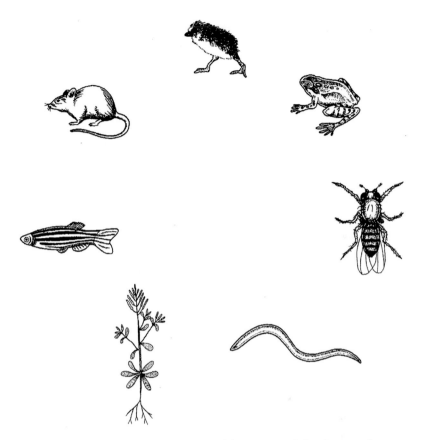

FIGURE 5 The seven main 'model systems' of developmental genetics. Clockwise from top: chick, frog, fruitfly, roundworm, thale cress, zebrafish, mouse.

However, the rationale underlying the question of which model system to choose for the study of development has shifted over the years, as the nature of the research concerned has changed. At the risk of offending historians of science by oversimplifying, it is approximately true that developmental biology has gone through four main phases. The first was the descriptive phase, where an embryo was simply observed and the details of what happened recorded. I say 'simply', but this is often not simple at all. What is needed to maximize progress is a creature where ideally we can see what happens as it happens in a living embryo or, if that fails, then we can at least compare the structure of embryos that died at different stages of their development and infer the temporal process from the differences we observe. In this early descriptive phase, the obvious first choice was the chick: cheap to acquire, big eggs, embryos visible through holes made carefully in the shell, rapid development, and a ready supply of more eggs if we need them. So it comes as no surprise that the earliest pioneers of embryology, such as Hieronymus Fabricius and Marcello Malpighi in the seventeenth century, studied the development of the chick.

The second phase of developmental biology was comparative. That is, it involved comparing the details of how embryos develop in different kinds of animal. Much of this comparative work was conducted by nineteenth-century embryologists such as Karl Ernst von Baer and Ernst Haeckel. I discussed this work in Chapter 2 and so will not elaborate here, except to note that because it was comparative work, it helped to usher in the use of further model systems (mostly vertebrate ones).

At about the start of the twentieth century, the third, experimental, phase of developmental biology began. The shift to a more experimental approach was led by the German embryologist Wilhelm Roux. For about fifty years or so this was the dominant *modus operandi* in studies of development. Some of the experiments involved were unsavoury to say the least. One line of investigation was to transplant part of a developing embryo to the wrong place and see what effect this had. Such studies, though gruesome, were very informative. They told us that some parts of an embryo had organizing effects on other parts, and induced those latter parts to develop in certain ways. A leading figure in this work was Hans Spemann,[1] and a particularly influential region of early-stage vertebrate embryos has been named, in his honour, as the Spemann Organizer. Experimental embryology made use of the classic model system of the past, the chick, but also newer ones. Amphibians are particularly useful material in this respect, with their large, conspicuous eggs lacking hard shells and thus facilitating experimental manipulation of the embryo.

The fourth, and now dominant, phase of developmental biology was/is the genetic one. This is hardly surprising; genetics seems to have hijacked most of today's biological research. Genes are involved in everything: the functioning of a single cell; the control of the life cycle; the transmission of traits from parents to offspring; and the evolution of one type of organism from another. But the creatures that are best suited to studying the involvement of genes in development are not necessarily the same as those that provided the best material for the earlier descriptive, comparative and experimental approaches, so this fourth phase has produced yet more model systems. The fruit-fly, roundworm, zebrafish and thale cress are all workhorses of this new era. Of course, the fruitfly had been used by geneticists from around 1900; but its developmental potential did not really emerge until about 1960, with the work of that American pioneer of developmental genetics, Ed Lewis.

* * * *

The challenge that I set for myself in the chapter title, namely to explain how bodies are built, now requires me to make two decisions: how many, and which, of the seven model systems should I concentrate on? It is my aim throughout this book to minimize technical detail, at least as far as is consistent with telling the story that I wish to tell. So, in line with this policy, I will choose just one of the seven. That's the first decision made. And with regard to the second, the decision of *which* one, I have little choice. I have spent a good part of my scientific life studying fruitflies and other arthropods – their development, their genetics, their ecology and their evolution – and it is always better to talk about a creature you know rather than one you do not. My knowledge of the other six model systems is all second-hand, unless you count my casual observations on tadpole development in my garden pond, and that kind of knowledge is never quite the same as knowledge that is at least partly gained from personal research. So the fruitfly it is.

Of course, I don't expect you to be all that interested in fruit-flies *per se*. To tell you the truth, neither am I. What we all want to know about are the secrets (which we hope are general ones) about how multicellular creatures manage to perform the various cellular and molecular tricks that are necessary to get through the whole life cycle from egg to adult and back again. But extreme caution is needed here. Since a fruitfly egg develops (eventually, via the larva) into a fruitfly whereas a frog egg develops (eventually, via the tadpole) into a frog, there must be aspects of the developmental processes that are very different if two such different end results are to ensue. So the real challenge for students of 'model systems' is to be able to distinguish the general from the specific. Some facts about fruitfly development apply equally to frogs and everything else; others are specific to fruit-flies. And there are facts of intermediate generality too, for instance those that extend from flies to moths and perhaps also to beetles, but not to much else.

I'm going to describe first an insect-specific aspect of fruitfly development and second a pan-animal one. The second story will be longer than the first, because generalization is my ultimate goal. But the first story is so bizarrely fascinating that I can't resist telling it, at least in brief.

* * * *

Picture a big fat maggot. Actually, fruitfly maggots are small and slim, but I always find it easier to picture interesting things happening in big creatures than small ones; and anyhow, housefly maggots, which are truly big and fat, exhibit very similar developmental processes to their fruitfly cousins. A fly life cycle consists of the following stages: egg; small larva (or maggot, if you prefer); bigger larva; pupa; adult. This is only a slight simplification in the sense that larvae grow by moulting, and there are usually at least three larval stages rather than just two. But no matter.

We are all familiar with the everyday miracle that the mag-got that ceases crawling around eating things and envelops its now-immobile form in a tough pupal casing is a very different creature from the winged adult fly that emerges a surprisingly short time later from a hole in that same tough casing. Familiarity doesn't breed

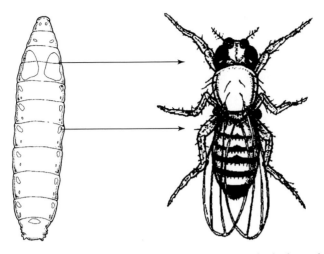

FIGURE 6 The special pieces of tissue inside the larva that, at metamorphosis, produce most of the adult fly.

contempt in this particular case – awe is a more appropriate emotion – and it doesn't breed mechanistic understanding either. What on earth is going on in there to turn one animal into another? If we didn't know better, we might venture 'magic' as our best attempt at an answer.

But magic it is certainly not. Rather, it is a rapid series of complicated developmental events. Here, in a nutshell, is the story. When a maggot hatches from its egg, it carries within it several tiny lumps of tissue called 'imaginal discs' that are invisible to the outside observer (Figure 6). Strangely, as the maggot grows, most of its constituent parts grow by an increase in the sizes of the cells of which they are composed rather than an increase in their number. But the tiny imaginal discs grow differently – here the growth is indeed achieved through increasing numbers of cells within each disc. So we already know that these lumps of tissue are special in some way. But just how special does not become apparent until later, when the maggot's urge to pupate finally gets the better of its urge to continue eating.

Inside the tough pupal casing, metamorphosis now takes place. But this does not involve gradually distorting the maggot shape into a fly shape with a minimum of fuss. Instead, many of the maggot's tissues are destroyed – broken down by its own enzymes. They are not used in the production of the fly. Rather, the cells in the small lumps of disc tissue now proliferate wildly, with a resulting increase in the sizes of these pieces of tissue, which lose their earlier quasi-disc shapes and begin to take on more precise forms. Different discs make different parts of the fly. There are wing discs, leg discs, etc. Hundreds of genes controlling thousands of cells produce one kind of creature from another in this bizarre way; and in a fruitfly kept in an incubator at the temperature of a hot summer's afternoon, the whole transformation takes just a few days.

Fascinating as the story is, its relevance outside the insect world is limited. Human embryos do not have imaginal discs buried deep within them; nor do frog embryos, despite the fact that they, like flies, are indirect developers and proceed to adulthood via metamorphosis. And even in the arthropod world, many groups – centipedes, for example – do not have equivalents of the fly's imaginal discs. That is why this particular process does not help us, at least on the face of it, to arrive at a general theory of how development works, regardless of the animal that emerges at the end of it. At a deeper level, though, it might help, because all those invisible happenings within each disc, such as the switching on of some genes in some places by the products of other genes, are indeed general, even though the discs themselves are not. However, to get at these happenings, I'm going to move back in time to when the maggot is still an embryo encased within its millimetre-diameter eggshell.

* * * *

Maggots, flies and all other arthropods have what is called a segmental body plan. That is, as you move along the head-to-tail axis, as you would if you were a tiny food particle in the gut, you pass through a series of quasi-autonomous segments. We humans and all our vertebrate cousins are also built on a segmental body plan, even though our skeleton, where this is most obvious (especially in the vertebrae that we're called after), is inside rather than out. And the members of a third major animal phylum, the annelid worms (such as earth-worms and lugworms), are also segmental in structure. Yet neither the eggs nor the earliest embryonic stages are segmented in any of these groups. So where do segments come from? That is, how are they made?

The story of how segments form in the embryo of a fruitfly has become one of the great classics of developmental genetics. It is to this field what *Macbeth* is to English Literature. If genes are the characters of this biological play, then there are about forty players. If you want to find out about all forty, read the book *The Making of a Fly*, by British biologist Peter Lawrence.[2] What I will try to do below is to give you a general insight into the nature of the plot with as little technical detail as I can manage. I can probably do this with reference to four genes rather than forty.

The very beginning of development (whether in flies, humans or any other creatures) poses a rather special problem. Suppose at 'time zero', say just after fertilization, the egg is a perfect sphere with a central nucleus surrounded by a homogeneous cytoplasm. How do we generate a heterogeneous embryo from such a homogeneous start? The same question can be posed a bit further down the line, if the egg undergoes repeated cell division to produce a ball of many identical daughter cells. How do we get from here to a highly structured embryo with head and tail ends, left and right sides?

Over the years, developmental biologists have come up with three different answers to this question. First, the homogeneous egg can use cues from the outside world to initiate directions for its body axes. For example, an embryonic plant can send its root downwards in response to gravity, or its shoot upwards in response to light. Second, the act of fertilization can itself be a symmetry-breaking event; the point of sperm entry can be used as a reference point to mark one end of some invisible axis around which development can subsequently organize itself. This is more arbitrary than using gravity, but it works for some creatures. Third, the concept of a totally homogeneous and spherical egg may be an illusion and the problem of giving the embryo information about, say, which end should become its head may have already been solved by the mother. This is what happens in the fruitfly, where the egg is ovoid and one end of it is chemically different from the other, with the result that some genes get switched on in the anterior end, others in the posterior end; this process provides some initial hooks on which to hang later refinements, of which there are many.

So, the genes that set up the initial chemical difference between the two ends of the egg have their base of action not in the egg itself but rather within the mother; and specifically, in the marvellously named nurse cells that surround, and pump things into, the egg while it still resides within the mother's reproductive tract. These 'things' include the product of a gene called *bicoid*, which is pumped into the prospective head end of the egg/embryo. This product remains at high concentration only at this end, and causes the activation of certain embryonic genes there but not elsewhere. One of these is *hunchback*. By the way, the naming of genes is a bit of a mess. All you probably need to know is: genes are often called after the mutant effects that arise when they go wrong; a sense of humour influences many gene names too (for example, in a different context, there is a gene called *Sonic Hedgehog*); many gene names are inscrutable unless you know a lot of background to their discovery; the names are often italicized to make them stand out, as you can already see; and the most important thing about each gene is not its name, but what it actually *does*, in terms of its effects on the embryo. This is what my skeletal coverage here will concentrate on.

What this first character – *bicoid* – in our much-simplified play achieves is the switching on of the first of the embryo's own genes that have a role in its development. We go from a head-to-tail gradient of a maternal gene product to a similar head-to-tail gradient in an embryonic gene product, simply because high levels of *bicoid* product are needed to switch on the *hunchback* gene. The embryo has taken over from its mother control of its own destiny.

But a head-to-tail gradient is a long way from a series of segments. This is where things get really clever. Suppose a gene gets switched on only at intermediate concentrations of the product of an earlier-acting gene. This is exactly what happens in the response of a gene called *Krüppel* to the gradient of *hunchback* product. This enables the embryo to make a 'stripe'. That is, a thin transverse band of tissue where the gene concerned is switched on, somewhere in the middle of the gradient of the earlier-acting gene's product. But once you have a stripe of gene expression it is easy to make more stripes involving other genes, if there are gene-switching mechanisms that give instructions like 'make a stripe of gene Y expression just anterior to the stripe of gene X', or alternatively 'make a stripe of gene Z expression in between stripes of X and Y'.

There are lots of such instructions to turn genes on or off in particular parts of the embryo, in certain spatial relationships to other, pre-existing peaks and troughs of gene activity. This is what I meant when I said that the very first heterogeneities that distinguish head from tail are like hooks upon which to hang later refinements. In the segmentation story, these refinements eventually produce one expression stripe per segment of a gene called *engrailed* (originally a heraldic term). The phrase 'per segment' is a bit misleading here because these *engrailed* stripes are one of the first signs of segmentation, so the segments as we can later see them in a maggot are not yet there. But in principle the building of all the structural detail of a segment can be hung on the hook of each stripe. So eventually, through a long series of steps, maternal gradients have generated embryonic gradients that have generated a few stripes that have generated more stripes that have generated other features of the segments that we can actually see with the naked eye.

What should we call this 'long series of steps'? I don't mean just this particular one, that produces segments, but rather all such series, whether the end products are segments, limbs, eyes or whatever. The three most commonly used terms for this are: developmental pathway, developmental cascade, or developmental hierarchy. None of these is perfect. 'Pathway' suggests linearity of interactions, whereas a complicated branching arrangement is the norm. 'Cascade' suggests one-way flow, whereas in some cases a gene, when switched on, can exert negative feedback on its own activating gene and switch that off. And 'hierarchy', given its origin, suggests a neat array of levels from the angelic master genes of earliest development to the priestly hordes of worker genes that put the final details in place; whereas the real situation is much more complex, with some genes being switched on again for a second burst of activity long after they had previously been switched off.

Given that it is hard to choose one of the three phrases over its equivalents in terms of accuracy – since they are all inaccurate – I will resort to my aesthetic preference for 'cascade'. As I said earlier, life flows; and what better to represent flow than a cascade? There is another advantage too. The relative positions of genes in these flows of cause and effect are often referred to as 'upstream' or 'downstream'; these connect perfectly with the idea of a cascade.

* * * *

At last it is time to say goodbye to maggots and to venture out into the exciting but dangerous waters of pan-animal generalization. What has the maggot, or for that matter the mouse or the worm, told us about the general principles about how eggs become embryos and embryos become adults? How are bodies, in general, built? How does each cell know what to do? Are we any further on in answering these formidable questions? Well, yes we are. Cells know what to do because genes tell them. (But this is a two-way process: genes know what to do because cells, through various signalling systems, tell them.) One cell will divide into two daughter cells while another will not because one has a particular gene switched on while the other has it switched off. And the pattern of some cells dividing and others not is what gives rise to the shape of the developing embryo. This is how bodies are built.

However, we should not let the numerous advances in developmental genetics make us become complacent. In a sense, the ways in which genes and signalling molecules interact, as revealed by the molecular work of the last two decades, constitute the 'chemistry' of development. But development also has a 'physics' (or a 'mathematics' if you prefer). In other words, developmental processes are systems with quantitative dynamics. This aspect of development is less advanced than the qualitative aspect of which molecules interact with which. However, pioneering efforts have been made by a few scientists, such as the German Hans Meinhardt[3] and the Canadian Richard Gordon.[4]

I have left many details of the story untold. This is hardly surprising, as I have allowed myself just one short chapter. For comparison, the latest editions of some of the leading developmental biology textbooks[5,6] have grown to more than 500 pages in order to accommodate the ever-increasing wealth of facts at our disposal that are the fruits of the last decade's research.

* * * *

Just as any particular developmental system can be thought of as a cause-and-effect cascade, so too can the whole of development. The building of a human, a chick, a fly, or any other creature of many cells from a simple unicellular beginning is a gigantic cascade – a cascade of cascades if you like. The repeatability of this grand cascade from one individual to another is truly remarkable. There is an almost infinite number of ways to arrange 100 trillion cells. Yet in us humans, who are of about this magnitude, the vast majority of these never come about. In an earlier book[7] – *Theories of Life* – I alluded to a ridiculous extreme body form: that of a 'linear human' in which all 100 trillion cells were arranged in a line. A crazy thought, perhaps, but it serves to make the point that most conceivable variations in the pattern of cell arrangement are never seen.

But equally, we are not all the same. Humans differ between families, between countries and between races. Even between siblings there can be big differences. I am about 5′ 6″ (1.68 m), my younger brother about 6′ 2″ (1.88 m). Why? He has brown eyes, mine are blue. Again, why? Largely because, although we have many genes in common, we also have many that are different. So you would hardly expect us to look like identical twins, where all the genes are the same.

Different traits get finalized at different times. It is easy to tell from a comparison of two babies that one is black and the other white. It is, in contrast, difficult to tell which one of two almost identical-sized babies will give rise to an adult that is significantly taller than the other. And hair colour at birth tells you almost nothing: my elder son was born with dark hair, turned into a toddler with blonde (almost white) hair, and now is a teenager with mid-coloured ('mousy') hair. Some differences between individuals would probably be visible by one or two months into pregnancy, if only we could see inside the womb. Other differences will wait until puberty or later to show themselves. I have two teeth more than the standard thirty-two (well, I did have before a few extractions) because two 'supernumerary' teeth decided that there was enough room behind my wisdom teeth to make an appearance. (They were wrong and have been rather troublesome.) This difference between me and someone born on the same day as me who ended up with the conventional thirty-two would not have been apparent until we were about the age of twenty.

* * * *

Now let's jump from development to its longer-term equivalent, evolution. Contrary to what most of the evolutionary trees displayed as posters in our museums or as prominent pictures in introductory biology texts might suggest, there is no way that an adult can evolve directly into another adult. The only way that natural selection can make new kinds of adult is by altering the course of development. And it can only make use of the variations that it finds at each stage of the life cycle. So some things can be modified in early development, others not until later. This interaction between natural selection and the different ways in which variation manifests itself at different stages of development determines where evolution ultimately takes us. Yes, *the interaction*, not one side or other of this interacting duo on its own.

One of the classic evolutionary case studies, alongside Darwin's finches and melanic moths, is the horse. The fossil record tells us that, among other changes, horses have increased dramatically in body size, starting from ancestors that were more similar in size to dogs than to the horses of today. But as usual the developmental side of the story is missing. Were those dog-sized horses different as early embryos from their present-day descendants? If we saw two horse embryos separated in time by millions of years, could we tell them apart? These questions cannot currently be answered, and perhaps they never will.

But no matter. My aim in the end is to tell you neither about maggots nor horses but about the general principles of the two kinds of biological creation and how they interrelate. So, the best strategy is to neglect whichever questions seem unanswerable and to concentrate on what the British biologist Peter Medawar[8] called *The Art of the Soluble*. So we continue, now, with some further questions that are answerable (at least in part), despite the fact that many of the creatures concerned lived millions of years before the horse was even a twinkle in its ancestor's eye.

5   A brief history of the last billion years

Bizarrely, it appears that space, time, energy and matter all erupted from a dimensionless point in nothingness about 15 billion years ago. So my 'brief history' covers the last one-fifteenth of the whole of time. It will, however, be a very biased – or do I mean selective? – history of that particular period. Bias and selection will, as you'll soon see, turn out to be two of the most important words in the book, and it won't hurt to examine their meanings a bit here. But this will be easier to do if I first tell you about the various ways in which my history is going to be biased or selective; let's just call it incomplete for the moment.

First of all, I will deal only with our own planet (age c.4.5 billion years), despite the fact that there are probably lots of other biospheres elsewhere with their own fascinating evolutionary processes being acted out right now, as you read this. Since we know nothing about these, apart from the fact that there don't seem to be any in our own solar system, I can hardly be blamed for this massive omission.

Second, I will deal almost exclusively with the history of *life forms* in my chosen period, and will say little about the physical world. This is an obvious choice, given my overall focus, though we should never forget that all creatures inhabit particular places, so the physical world matters to them, in both the short and the long term. The ambient temperature can deflect development one way or another, as when it determines the sex of a turtle; and continental drift can alter the course of evolution. So the physical world is important to both of the great processes of biological creation.

Third, I have a bias towards the animal kingdom, due to my own 'academic upbringing'. However, I am confident that the relationship between development and evolution that we are beginning to discern in multicellular creatures applies to animals, plants and fungi. No doubt the details will turn out to differ somewhat between kingdoms, but not, I hope, the general principles.

Finally, the history of life is, as they say, written in the rocks. Our only direct source of information on the organisms of the past are the fossils of the present. But the fossil record is itself biased in many ways. Whether or not a dead creature fossilizes depends on all manner of things: its location, the substrate, the weather, the presence of scavengers and so on. The tiny fraction of all creatures whose bodies (or parts thereof) we find fossilized is thus a highly biased sample of the overall fauna, and an accurate quantitative estimate of this bias is, if we are honest, beyond our reach.

So, back to the question of whether these restrictions to my history are sensible selections or bad biases. I have deliberately put it in that way to highlight the fact that exercising choice (i.e. making a selection) is normally thought of as appropriate in many situations, while exercising a bias has, in most contexts, a disreputable air. I choose to watch one football match rather than another (sensible; I want to watch my own team play, not teams in which I have little interest); but I get annoyed, like most folk, if the referee seems to be biased. Choice, or selection, implies an unspecified range of options from which one is chosen on the basis of some rational criterion. Bias, on the other hand, implies violation of some desired even-handedness between two (or sometimes more) equally deserving causes. For the moment, I will accept this distinction, though I will revisit it in Chapter 8, where I will make a passionate argument for 'developmental bias' being a crucially important evolutionary mechanism that has been woefully neglected by mainstream theory.

Given the above distinction between selection and bias, it is now clear how the various restrictions to my history should be classified. Restriction to our own biosphere is a sensible choice. So too, in a book about life, is my decision to concentrate on the biological world rather than its physical counterpart. But the other two restrictions are biases, one deriving from limitations to my personal knowledge of plants and fungi, the other from the many factors affecting the probability of fossilization of dead creatures.

So far, I have ignored the most obvious restriction, the one that is embodied in the chapter title: a restriction to the last billion years, in other words less than a quarter of the Earth's history. The reason for this restriction is that this is the best ball-park figure that can be given for the age of the animal kingdom. If we choose to ignore all indirect information on this issue and believe only in what the fossils seem to be saying, we might opt for a later birth of animals, maybe around 600 or 700 million years ago. But if we listen to what the new generation of comparative molecular biologists are telling us, we might opt for an earlier birth – perhaps 1200 million years ago or even more. So my ball-park billion is a rough and ready compromise between these two extremes.

*　　*　　*　　*

You might ask why we should bother with history at all. Some branches of evolutionary biology, notably population genetics, seem to get by with very little history. My own doctoral training exemplifies this. Although I did my Ph.D. in an excellent department in a reputable university, and had an outstanding supervisor (Bryan Clarke, famous for his work on snails and his antagonism to Kimura's 'neutral' theory of molecular evolution),[1] I emerged at the end of my three-year tunnel clueless about the history of the biosphere. I had heard of the Cambrian period (more on which later), but I couldn't have told you when it started or finished. I knew there had been mass extinctions, like the one that killed off the dinosaurs, but I didn't know how many. Nor was I uniquely ignorant of these and other historical details. Most of my fellow Ph.D. students were similarly ill-informed about these things, though there were, as ever, occasional exceptions.

Don't take this as a belated attack on the quality of my education; far from it. The reason that I did not know these things was that neither I, nor my teachers, nor the authors of the books on population genetics that I read, thought they were important. We were united in our belief that what we were doing was investigating an evolutionary

mechanism – natural selection – that would operate in a broadly similar manner regardless of time or place. And perhaps we were right. It
may well be that in some remote solar system in the great galaxy of
Andromeda there is an evolutionary process going on that is driven by
natural selection, regardless of the fact that all the contextual details
(and the creatures) are different.

However, population genetics, if undertaken in isolation from
other branches of evolutionary biology, can all too easily turn into
playing mental games with algebraic equations in a sterile void.
Despite the potentially general applicability of natural selection to
biospheres throughout the universe, we cannot neglect that we are
ourselves firmly rooted here on planet Earth, linked by an unbroken
chain of animal ancestors to the unicellular world of a billion years
ago. Given that fact, we can hardly ignore the historical dimension of
our own evolutionary process – the details of what evolved into what,
and when and where such transitions happened. Algebraic equations
should not be thrown out of the window; they should be enriched
with historical information.

Now that I have evolved from a population geneticist of sorts
into a student of evo-devo, I find the need for history compelling; hence
the following. But I have never lost my central focus on mechanisms.
So this book will be very different in its emphasis from one written
by a student of evo-devo who arrived in this strange hybrid discipline
from a route that started in palaeontology.

*   *   *   *

It makes sense to divide animal history over the last billion years into
three time periods. First, there is a period of about 300 MY (I'll use that
from now on for 'millions of years') up to about 700 MYA (A stands for
ago) for which we have no indisputable fossil evidence that there were
any animals at all. Comparisons of the genes of present-day animals
suggest that animals were around during this period, though whether
this kind of extrapolation into the distant past is valid remains to
be seen. Some biologists regard it as fine; others, including myself,
still have reservations. However, there is another very different kind

of indirect evidence for the existence of animals in this first 300 MY period. This evidence comes in the form of 'trace fossils', meaning not the fossilized remains of the animals themselves but rather fossilized remains of their activities, for example fossilized worm tunnels. The problem with these is that we have to be very careful in distinguishing worm tunnels from tunnels made by, for example, water running through sand just as it is being compacted into sandstone. In the case of true fossils, like dinosaur skeletons, the idea that they may be of non-biological origin is nothing more than a creationist's dream; but in the case of trace fossils, a chemical rather than biological origin often remains a possibility.

So we should be honest and admit that for the period from about 1000 to about 700 MYA we know nothing for certain about the animals that were around. We don't even know if any *were* around, but it seems likely. The lack of a decent fossil record from this period is probably due to the animals concerned being very small and lacking hard parts, and so having an exceedingly low probability of leaving fossilized remains. Time will tell if this conjecture is correct; new fossils come to light every year.

The second of my three phases of animal history extends from about 700 MYA until 543 MYA. You can immediately see from the precise (and possibly even accurate) end year that we are getting into the realms of more definite knowledge. This figure of 543 MYA represents the beginning of the Cambrian Period, and since my non-palaeontological readers will probably find this and other official period names as hard to remember as I do, in the sense of connecting them with absolute time, I have set them out in visual form for easy reference in Figure 7. Incidentally, many period names derive from places where fossils of the appropriate age were found. The Cambrian is named after Wales (Cymru in Welsh). I recently visited my brother who teaches at a branch of the University of Wales in the picturesque little town of Lampeter, and noticed to my amusement that the local newspaper was called the *Cambrian News*. I didn't find any articles in it about fossils.

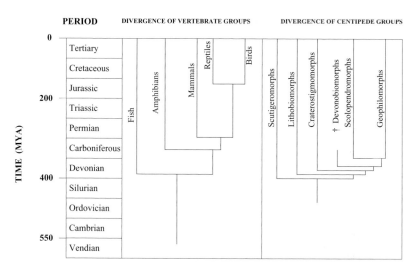

FIGURE 7 The geological periods from 600 MYA to the present (left) and the patterns of divergence of groups of vertebrates (centre) and centipedes (right). Timescale is non-linear and very approximate.

The largest collection of fossils from this phase of history – from 700 to 543 MYA – represents a group of animals called the Vendozoa (after the period immediately before the Cambrian, which is called the Vendian). These were first found in the Ediacaran Hills in Australia, and so are also sometimes referred to as the Ediacaran fauna. However, this is a potentially misleading term because these creatures were not restricted to Australia. It is now apparent from numerous fossil finds that this was a world-wide fauna that persisted for many millions of years.

It is by no means clear what these creatures were. They took a variety of forms, some of which are shown in Figure 8. Most biologists believe that the Vendozoans were indeed animals, though there is not even complete agreement on that. There is less agreement as to what kind of animals they were. They may have been early representatives of some of the animal phyla that we recognize today; some may, for example, have been close relatives of jellyfish (phylum Cnidaria). Alternatively, they may have been an entirely distinct evolutionary radiation of a type of animal body plan that disappeared from the

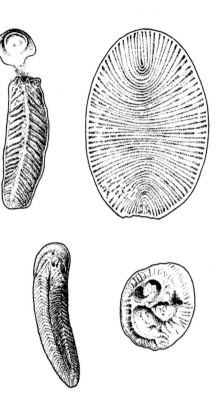

FIGURE 8 A selection of fossil Vendozoans.

face of the Earth somewhere around the beginning of the Cambrian, a view that has been championed by the German palaeontologist Dolf Seilacher.[2] They may even have been a mixture of the two, given the wide range of different kinds of Vendozoan fossil.

I have to declare myself an agnostic in relation to this issue, though perhaps an agnostic with an inclination to believe the 'mixture' hypothesis. In any event, it appears from the widespread Vendozoan fossils that there was a well-developed fauna at that time, but that even if some of today's phyla, such as Cnidaria, were present, others were almost certainly not. None of the Vendozoans looks like a vertebrate, for example, and given that conditions in the Vendian were clearly suitable for wide-scale fossilization, it seems unlikely that creatures with well-developed skeletons, if present, would have left no trace.

So, on to the third and final of my three phases of animal history: from 543 MYA to today. This enormous span of time is referred to as the Phanerozoic Era, which literally translates as 'manifest animals'. This label was given by early palaeontologists who knew nothing of the Vendozoa, and thought that no animals existed before the Cambrian. There is no way I can do justice to the animal history of the Phanerozoic here. So, I will do my usual thing: give some of the most important points, and illustrate them with a couple of examples.

The Phanerozoic began with one of the most dramatic events in the history of the planet: the 'Cambrian explosion'. Suddenly (geologically speaking), almost all of today's animal phyla whose body plans render their fossilization potential non-negligible appeared all at once. There were Cambrian molluscs, Cambrian annelids, Cambrian arthropods, Cambrian chordates (including vertebrates), and so on.

The correct interpretation of this great explosion is unclear. It may have been genuine, in the sense that it was an explosion of evolutionary activity, with lots of new kinds of animals coming into being, their lineages rapidly diverging from each other as they all ploughed their new furrows into the morphological unknown. Alternatively, it may be a fake explosion in the sense that it was merely an explosion of fossilization of animals that were already in existence but had previously left no trace, perhaps because of very small body size. The comparisons of the gene sequences of extant animals that I mentioned earlier suggest the latter, but I'm not so sure. Perhaps this is another instance in which it is best to remain, at least for the moment, agnostic.

* * * *

You can begin to see a recurrent theme here; and it will recur again before we are finished. Do evolutionary events take the form of explosions followed by long periods of relative quiet, or are they, rather, plodding things that vary a bit but not a lot in their rate, and are falsely made to seem explosive by artefacts of the fossil record? This question has been asked not only in relation to events going on at different points in evolutionary time, like the Vendian and Cambrian, but

also in relation to what you might call different levels of evolutionary change. In considering the Cambrian explosion we are focusing
on a high level of evolutionary divergence – that of different phyla.
The American palaeontologist G. G. Simpson called this the mega-
evolutionary level. When Niles Eldredge and Stephen J. Gould introduced the theory of punctuated equilibrium[3] in 1972, they were dealing with the same issue but at a lower level – that of the evolution of
species and genera, which is usually referred to as the macroevolutionary level (leaving 'microevolution' to refer to changes going on within
the confines of a species, for example as it spreads to new areas and
adapts in almost imperceptible ways to the local conditions in each).

This issue of explosions versus plodding is hugely important.
However, it is also rather dangerous – people are apt to take one side
or the other because of their general ideological outlook rather than a
dispassionate consideration of the facts inasmuch as we know them.
No one is immune to this danger; the completely objective scientist
is a myth. I find myself drawn more to explosions than to plodding,
though I suspect that evolution involves plenty of both.

By the end of the Cambrian, all animal phyla were probably represented among the fauna of the time. (Why have no phyla originated
since? This is an interesting but as yet unanswerable question.) So
animal history from then to now can be regarded as a multitude of
refinements, both large and small, on the thirty-five or so body plans
that were established in the Cambrian or perhaps earlier. But clearly
this overall multitude cannot be dealt with in a book of this kind,
so I will restrict myself to two illustrative stories, one familiar, the
other probably not: vertebrates (50 000 species) and centipedes (3000
species) respectively.

We all know the vertebrate story, at least in outline, from visits
to museums or from natural history programmes on TV. So I will be
brief. The first vertebrates were fish – marine rather than freshwater,
of course; most evolutionary tales begin, ultimately, in the sea. The
word 'fish' covers a great variety of creatures of considerable structural
disparity – for example skeletons made out of cartilage versus those

made out of bone – but I will stick with the traditional bias of narrators of this story and ignore most fish from here on and concentrate on what happened when one line of fish began to invade the land. The pattern of divergence of the main groups of land vertebrates is now well established. The fin-to-limb transition resulted in early amphibian-like 'tetrapods'; strangely, it now seems that these were still aquatic creatures, though their descendants eventually became partly terrestrial, and led the way to today's amphibians. Subsequently, evolution of greater independence from water led to the reptiles. Later on, two different reptile lineages independently evolved into the two 'warm-blooded' vertebrate groups – mammals and birds. It is now pretty clear that birds derived from a lineage of dinosaurs. Mammals had a separate reptile origin in the group known as the synapsids. The approximate times of these various divergences are shown in Figure 7.

Now to centipedes. Currently the world has about 3000 different species of centipede that we know about, and maybe a similar number that we don't, most of which are doubtless lurking in those poorly explored biodiversity hotspots the tropical rainforests (or what's left of them). And, as with any other major group of animals, there must have been countless species, now extinct, only a tiny fraction of which have been kind enough to leave us some fossilized remains.

So what of centipede evolution? What do we know about how the centipedes of today arose from the centipedes (and proto-centipedes) of the past? It seems pretty clear that there were no centipedes in the early Cambrian, because all animal life was then still marine, and there are not, and probably never have been, any ocean-going centipedes. The whole centipede body plan is very much geared to life on land, as is that of its myriapod relative the millipede. The earliest fossil records of any kind of myriapod derive from the Silurian Period (see Figure 7). From their general appearance, though, they were not yet centipedes. They were clearly arthropods with many segments and many pairs of legs, but they lacked the centipede's distinctive and lethal weapon – its pair of forcipules or poison claws. It is unclear who the closest relations of these earliest myriapods were. The old story of a close relationship between myriapods and insects has died as a result of various recent studies, but no particularly clear alternative story has yet replaced it. So all we know is that the first myriapods arose from some non-myriapod arthropod lineage, but which one remains to be seen.

If we come forward into the Devonian, there were definitely centipedes, but some of these were of a kind that must have subsequently become extinct, because there is none of them among today's fauna. However, if we come forward again to about 300 MYA, to the Carboniferous rocks of the Mazon Creek formation in Illinois, we find the remains of creatures that not only were centipedes but bore a remarkable resemblance to some of those that we see around us today.[4] These fossils belonged to the group called scolopendromorphs that includes today's most lethal, foot-long tropical centipedes that are definitely best avoided by all but the most intrepid centipede-hunters.

It is probably true to say that all the major types of centipede (there are six of these altogether) had come into existence by 300 MYA, though their fossil record is so sparse that we cannot be sure, and one group, the long thin subterranean geophilomorphs (literally 'ground-loving') could conceivably have come on the scene a bit later. Now 300 MY is a long time to do nothing. I don't mean quite literally nothing, of course, but nothing in the sense of inventing major new subdesigns of the centipede body plan. So perhaps this is an example of the 'explosion followed by stasis' pattern. But we really can't be sure because there are so few centipede fossils. Yet another tantalizing evolutionary story that we lack the evidence to complete.

These tales of vertebrates and centipedes are very different, and yet in another way very similar. They are different because the actual animals have very different structures; that is, evolution has taken very different routes in the two cases. The fangs of a snake and the poison claws of a centipede are both pairs of dangerous sharp appendages at the anterior end of the body, but this is only a superficial similarity. There is no direct evolutionary relationship between the two; or, to put it another way, they are not homologous structures. Nevertheless, the evolution of vertebrates and centipedes is, at a more general level, the same. In both cases, many lineages all characterized by possession of a common body plan radiate out from each other as they invent new variations of the plan concerned. These inventions involve changes in the relevant developmental cascades. This can be stated with certainty because, although we have little fossil evidence of embryos, we know that altering the course of development is the only way to produce, in evolution, a different kind of adult structure. Even in the case of unusual genetic mechanisms underlying evolutionary change – such as lateral gene transfer – this point remains valid.

What is true of vertebrates and centipedes is also, at this level, true of the whole animal kingdom, and indeed of plants and fungi too. That is why I do not, even in this historical chapter, have to give you an exhaustive list of what has happened during the evolutionary history of every different kind of creature. Anyhow, if you are desirous of further stories of this kind about other groups of animals, or about plants and fungi, there are many good books to choose from. And you can't expect a former population geneticist to be a well-rounded historian of the living world; it's much better to get a detailed history, if you want one, from a palaeontologist.

* * * *

So, enough of history. We are now about to return to mechanisms. In all evolving lineages of multicellular creatures, from the pre-Cambrian to the present, evolutionary change has happened because developmental cascades have somehow been modified. The results of this modification process are many and varied: segments, legs, feathers, flowers, shells, eyes, teeth, brains. How have all these, and many other commonplace features of today's creatures, come about from a featureless beginning in the unicellular world of a billion years ago? Charles Darwin came up with a major part of the answer, namely natural selection. But are we content to believe that this is the whole answer? I, for one, am not.

This question brings us back full circle to a point that emerged in Chapter 3. I said then that development could be brought into the evolutionary synthesis in two ways, one of which was controversial, the other not. The latter is no problem and I need not dwell on it. No reasonable person is going to object to me or any other biologist saying that our theory of evolution would be more complete if it took on board not just the selective reasons for a bird's beak getting bigger but also the nature of the developmental system that produced the ancestor's particular beak and the ways in which it got modified to produce the different beak possessed by its descendant. But many people may object if I say, as I will, that these very developmental changes are in part responsible for the directions that evolution takes and that natural selection in response to environmental factors is not the sole mechanism underlying evolutionary directionality.

6   Preamble to the quiet revolution

The starting point for my preamble to the book's core chapters (7–11) is an attempt to distinguish between different hierarchical levels in the battle against ignorance. At the most general level, all humans of goodwill are fighting this fight, whether by actions or by words. A student on a year out doing voluntary work in a deprived region of Africa is fighting ignorance regardless of whether he or she is teaching in a classroom or digging irrigation ditches. I make this point at the outset because I do not wish to be seen as proclaiming that academics have a monopoly in this particular battle. But having made it, I will indeed focus down one level in the hierarchy and restrict myself to the academic domain.

What is it that characterizes academic enquiry as a particular branch of the battle against ignorance? This is a remarkably difficult question to answer. The fact that I know this is one of the few benefits that derive from membership of high-level university committees. Across a university, academics study a wide range of disciplines: from art to philosophy, from business to biology. The ways in which they study may have some things in common, but in many inter-subject comparisons the differences are more conspicuous than the commonalities.

It is the 'way' or 'method' of study that distinguishes the scientific disciplines from others. Again, as we descend another level in the hierarchy, and so are faced with a choice, I will take the direction that leads to the domain about which I am better informed rather than the alternative one. So, we are now headed for 'science' as opposed to 'non-science', or everything else. A highly biased split, perhaps, but I leave it to others who are better qualified for the task to decide how the 'everything else' domain of academia should be divided in terms of methods of study.

Well, I am now more comfortable because I have reached a level at which I feel at home. Although my knowledge of the non-biological sciences, whether chemistry or cosmology, is rudimentary when compared with that of specialists in such fields, I share with them the so-called 'scientific method'. However, defining this is something of a minefield, and one in which the mines are laid not just by other scientists but by philosophers too. What follows is a very personal account.

*   *   *   *

The scientific method is not a singular thing. I want, in particular, to distinguish two aspects – you might even say levels – of this 'method'. We'll start with what seems to me the lower, more workmanlike, level. This is the level of the specific question and the individual experiment. For example, suppose I want to find out whether a particular chemical can act as a fertilizer that enhances the growth of a particular species of plant, and I am planning an experiment to test this idea (or 'hypothesis'). The experiment in this case can be very simple. I grow some plants in containers with some standard soil, and others in identical containers with identical soil, but in this second batch the chemical is also added. After a time interval that is long enough for substantial growth to occur, I measure the weights of the two groups of plants and see if they are different. If they are, then I have a positive answer to my hypothesis, which I might even be tempted to call a 'fact' – that is, chemical X has a positive (or negative) effect on the growth of plant species Y.

There are, of course, all sorts of complications, even in a simple experiment like this, and books in the field of experimental design deal thoroughly with them. I will be content here just to raise a few unanswered questions. How many plants should be used? What concentration of the chemical? Where should the experiment be done? How should the pots of 'experimental' (plus fertilizer) and 'control' (minus fertilizer) be arranged to avoid spurious results, as might happen if the chemical was really inert but the plants to which it has been added are all closest to the window? How big does the difference have to be before it is clear that it is a real difference? Which statistical techniques should we use to analyse the data?

I could just as easily have chosen other hypotheses and experiments from other fields of science and the general method would still apply. It is sometimes called the hypothetico-deductive method, or the Popperian method, after the philosopher Karl Popper who did much to elucidate it. But importantly, this is all small-scale stuff. What happens when we broaden our view? Do major scientific advances occur in the same way as the generation of individual scientific 'facts'? To say that this is a loaded question is an understatement *par excellence.* Different scientists, and different philosophers of science, do not agree on the answer. Indeed the answer may well differ between cases – 'yes' for some major discoveries but 'no' for others.

This issue has been addressed by another famous philosopher of science, Thomas Kuhn,[1] in his book *The Structure of Scientific Revolutions.* Kuhn took the (controversial) view that the kind of experiment that I described above, and others like it, belong to what he called 'normal science'; that is, the kind of science that proceeds all over the world on a day-to-day basis. He contrasted this with scientific 'revolutions' where, rather than generating the 278th 'fact' in an accumulating pile of such facts without really changing how anyone looks at the world, something special happens that causes a major shift in our world view. I will now focus on just two aspects of how Kuhn characterized these scientific revolutions.

First, although a revolution may reveal that the previously held view was wrong (as when Copernicus overturned the Earth-centred view of the universe that had prevailed, at least in the West, since Ptolemy more than a thousand years earlier), revolutions do not always take this particular form. Rather, they often take the form that the new world view recognizes the old as still being correct, but only under certain circumstances. That is, the new subsumes the old as a special case. The most famous example of this is Einstein's revolution in physics. We do not, post-Einstein, say that Isaac Newton's laws of motion were wrong. Instead, we recognize that they are right, but only when the bodies concerned are travelling quite slowly. If these bodies approach the speed of light, Newtonian predictions don't work any more because Newton's laws were not general enough to encompass such unfamiliar situations. Einstein's predictions, in contrast, do work, even if we non-mathematicians find them bizarre and counterintuitive.

Second, Kuhn pointed out that, due to many scientists' reluctance to shift their world view, especially a long-held one, revolutions sometimes gain acceptance in the scientific community not by people changing their minds but rather by the cohort of scientists holding the old view slowly dying out (or, less drastically, retiring), and being replaced by a new generation that is more receptive to new ideas because of having been immersed in the old ways of thinking for a shorter time. I think there is much truth in this, though as ever we need to acknowledge the existence of between-individual variation, and specifically in this case the remarkable honesty and flexibility that enable some elderly scientists to agree that the world view that has underlain their life's work has turned out to be wrong.

The 'subsuming' type of revolution (as exemplified by Einstein) is of particular interest because it tells us something of the utmost importance about what science is trying to do. It is trying to generalize. That is, it is trying to come up with 'theories' that subsume as many as possible individual cases. That is what makes it important. If Darwin's natural selection applied only to badgers it would hardly have caused bishops any great loss of sleep. But it doesn't just apply to badgers. It applies to all species on the face of the Earth today, as well as all their ancestors and all their descendants. As I noted earlier, it may apply also in non-terrestrial biospheres if such exist. So it has huge generality. That makes it a very important theory. And I hope you'll note the difference between this use of 'theory', which is a very positive one, and the creationists' frequent claim that evolution is 'just a theory' (meaning untested hypothesis). In my usage, and that of many others, evolution by natural selection is proud to be a theory, for this is a badge of distinction, not a cause for mistrust.

* * * *

We have now, without really trying, come down two more levels in the hierarchy that I have used to characterize the battle against ignorance: from science to biological science to evolutionary biology. So we are getting close to the task at hand – the shift in world view that I believe the advent of 'evo-devo' warrants. Now we have to ask an interesting question: which of the two types of scientific revolution just described – the wrong to right or the less general to more general – does it belong to?

Strangely, I think the correct answer to this question is 'neither'. It certainly does not belong to the former. I do not believe that Darwin was wrong. Natural selection may be a theory in the positive sense of that term, but it is certainly not an untested hypothesis. Many studies on many organisms have demonstrated its reality. I have no ambitions to suggest that Darwin was a sort of biological Ptolemy, and in this I am in full agreement with the vast majority of biologists.

Nor will a new and expanded evolutionary theory, in my opinion, subsume natural selection as a special case of something more general, in the way that Einstein subsumed Newton. However, I think that the change in world view that is beginning to take place in Evolutionary Biology, inspired by the advent of evo-devo, is a kind of variant on this second type of scientific revolution. This requires a bit of explanation, and I will make use of an analogy for this purpose.

This analogy involves pillars and arches. The Hungarian-born British writer Arthur Koestler, in his beautifully titled book[2] *The Ghost in the Machine*, described Darwinian natural selection as one of the four great 'pillars of unwisdom'. Now it will be apparent to you from all that has gone before that I believe Koestler was completely wrong in making this pronouncement. Natural selection is, in my view, quite the opposite of what Koestler considered it; that is, it is a pillar of wisdom. To be more specific, it is a pillar of scientific wisdom,

Mid twentieth century reality   Twenty first century goal

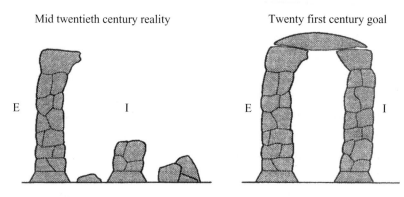

FIGURE 9 Evolution theory as an archway, one of whose pillars requires more attention than the other in the twenty-first century. E, external; I, internal.

and, going more specific again, it is a pillar of our understanding of how evolution works, and thus of how we and all our fellow creatures came to inhabit our planet.

But my problem is this. I believe that a complete theory of evolution is akin to an archway in that it will only stand firmly if supported by two pillars – in a sense an 'external' one and an 'internal' one (Figure 9). Darwin built one of these pillars – the external one concerning the adaptation of organisms to their environment. Although he provided a few bricks for the other internal one – his comments on embryology in *The Origin of Species* are characteristically astute – he did not build it to an equivalent height to his towering pillar of natural selection. This is the job that now confronts us, and is, in my view, well under way but still by no means complete.

I have previously – in *The Origin of Animal Body Plans* – described neo-Darwinism as lop-sided.[3] The pillar-and-arch analogy is, I hope, a good way to provide a vivid mental picture of this lop-sided condition. And this is a condition that has grown worse over time, at least between the late nineteenth century, when Ernst Haeckel was doing his own form of embryological Darwinism, and 1977, the year in which Stephen J. Gould published *Ontogeny and Phylogeny*.[4] This book did much to rekindle interest in the relationship between development and evolution – a rekindling that grew into an explosion with the discovery of the homeobox just a few years later.

The descent into greater lop-sidedness is, in my view, the result of the direction that the discipline of population genetics took as it was being born in the 1920s and 1930s. If you read R. A. Fisher's *Genetical Theory of Natural Selection*,[5] you won't find astute comments about development equivalent to those that can be found in *The Origin of Species*; in fact, you'd be hard put to find any comments about development at all. And this is despite the fact that much more embryology was known in 1930 than in 1859. In single-mindedly pursuing a specific goal – the 'mathematicizing' of natural selection – Fisher and his colleagues were unwittingly exaggerating evolutionary theory's lop-sidedness. I could hardly blame them for adding to the already-bigger pillar. They did a good job on this. But I can and do blame them for failing to see the obvious, namely that this pillar on its own would never produce a complete and sufficient theory of evolution.

* * * *

So, from 1930 to 1977 mainstream evolutionary theory became more and more lop-sided. The 'modern synthesis' may indeed be a synthesis of several previously separate strands, but it was, and is, only a partial synthesis. It is therefore also a weak synthesis. It is like an archway with one pillar missing. Thankfully, there have always been mavericks in science, and so there is always an alternative to the mainstream. During the period of about half a century when the 'synthesis' dominated mainstream thought, a few biologists ploughed separate and enlightened furrows. The two that spring most readily to mind in this period are the Edinburgh-based geneticist C. H. Waddington and his Moscow-based counterpart, I. I. Schmalhausen. If nineteenth-century figures like von Baer, Haeckel and Geoffroy Saint-Hilaire are thought of as the grandfathers of evo-devo, then Waddington and Schmalhausen must surely be considered as the next generation in this noble tradition, and we who labour in this field today the third generation. And with luck, the gaps between 'generations' are ever-decreasing.

But mavericks often have less impact than they merit on the course of human thought; and that is the case here. All the way to 1977, and perhaps a little beyond, mainstream evolutionary theory trundled on, with most of its proponents ignoring the mavericks. Neglect is much more efficient than criticism at ensuring that a scientist's work makes little impact. As the German taxonomist Willi Hennig[6] said, 'it is not worth expending critical energy on unimportant work'.

With the advent of present-day evo-devo, neglect is no longer an option. And acknowledgement of a role for 'constraint' is not enough, as I argued earlier. The key point that I am championing herein, and hope to persuade you to accept, is that development has a much more far-reaching effect in determining the direction of evolutionary change than merely closing off a few avenues. I believe that it does this in two main ways, one concerning the structure of developmental variation (Chapters 8 and 9) and one concerning the developmental system's integration as being itself a target of selection (Chapters 10 and 11). These two emphases are certainly assaults on the modern synthesis, because it rejects the first outright and de-emphasizes the second to a point where it is effectively damned with faint praise.

Few authors have made successful efforts to champion these ideas. As noted in Chapter 2, the orthogeneticists of the late nineteenth and early twentieth centuries argued that internal forces drove evolution in particular directions. But they had no mechanistic explanations for their proposals, and were therefore perilously close to the interface between science and mysticism. In the 1960s, one author, the Englishman Lancelot Law Whyte,[7] proposed a mechanism that he called internal selection. I believe that he was largely correct and I will expand on this belief in Chapters 10 and 11. Very recently, a pair of American authors – Lev Yampolsky and Arlin Stoltzfus – proposed a mechanism for biased introduction of variation[8] that I believe is entirely correct and will return to in Chapters 8 and 9.

You may have noticed, if you have a numerical leaning, that this is Chapter 6 and I keep mentioning Chapters 8, 9, 10 and 11.

What of Chapter 7? Well, as you will see shortly, its role is to deal with a problem that logically precedes all others in my endeavour: the problem that even the scientific language is stacked against what I am trying to achieve. A picture is indeed often worth a thousand words, but if a core word is entirely absent from our vocabulary then no amount of pictures or clumsy concoctions of multi-word phrases will plug the gap. So now we go mentally into a domain where many of our ancestors have gone physically – the domain of 'reprogramming'.

7   The return of the organism

Science is, as we have seen, all about generalizing. So my title refers to the return of 'the organism' in a general rather than a specific sense. And it refers to a return to what I consider to be its proper, central, place in the theory of evolution. We have already examined its initial *displacement* back in the 1930s. Now we need to take on the challenge of its *replacement* to a centre-stage position in the twenty-first century. This challenge involves correcting the situation that has arisen in which the gene and the population have, whether by design or by accident, combined to squash the organism out of evolution theory's core.

Let's begin with the 'selfish gene' concept, as championed by the Oxford-based biologist Richard Dawkins.[1] My feeling about this concept is that it is useful in one specific way but that its importance has been vastly overstated. Its usefulness arises from its ability to quash the naïve notion that evolution necessarily works 'for the good of the species'. Sometimes it does, sometimes not. It all depends on whether the interests of the species and the consequences of selection on organisms and families coincide. My favourite example of non-coincidence is this. A population of flies is growing rapidly because every adult female produces 200 eggs that, after larval mortality, become 20 adults. It is in danger of running out of resources several generations hence and thus becoming extinct. It would be beneficial to the population, and the species, if a mutant fly that appeared and was characterized by a reduced fecundity of 20 eggs could spread through the population so that all flies were of that kind. If the same 90 per cent mortality applied to such a 20-egg-per-female population, it would be the ultimate in what has become known, in more human circles, as sustainable development: two flies produce 20 eggs which produce two flies, and so on indefinitely.

But unfortunately, natural selection does not work in this benevolent way. If all else is equal, a variant laying 20 eggs rather than 200 will die out rapidly from the population, which will later go extinct as a result of its own short-sighted overuse of resources. In this case, natural selection acting at the level of the organism/gene has acted against the interests of the species. Of course, this is all just mental game-playing, and if we give the game different rules we will get a different result. If the 20-egg producer makes eggs that somehow are 'better' and have a higher probability of survival to adulthood, then the form of selection will change, and a lot hinges on how much better is 'better'.

The idea of the selfish gene really comes into its own when the interests of the gene and the organism part company. There are clearly cases in which this happens, and one of the best examples is that of the evolution of insect societies, primarily in the group of insects known as the Hymenoptera (ants, wasps and bees). In some hymenopteran species, as is well known, there are different social 'castes', including sterile workers. How can an evolutionary process that is based on the spread of fitter variants produce organisms whose fitness, by definition, is zero? Well, if you work through the appropriate equations (and they're really quite simple ones), it turns out that a gene that is bad for its bearer will nevertheless spread through a population by natural selection if it benefits close relatives by a degree that outweighs its primary problem. One of the factors that affects this process is the degree of relationship between relatives; and it is no accident that the genetic system that many hymenopterans possess results in their having an unusually high degree of relatedness and so renders this kind of selection – called kin selection – more likely.

So, there are specific cases where the linked concepts of the selfish gene and kin selection are essential tools to understand what is going on. My reservation about these concepts is that they seem to have become regarded – at least in some circles – as having a *generally* important evolutionary role. OK, there are many kinds of social hymenopterans, so we are not talking about something that is unique to a single species. But, over the whole span of evolutionary time, across the whole range of creatures, in what proportion of cases do the interests of the gene and the organism part company in this way? I'm not foolish enough to pluck a particular figure out of the air, but my gut feeling is that the correct answer is 'a small proportion'.

* * * *

At this stage, we pose the important question: how do evolutionary novelties (like heads) arise? So far, I have said that neither species selection (Chapter 2) nor selfish genes provide the answer. I have also said that Darwin provided part of the answer and we need to build on (or rather build beside, if you accept the twin pillars analogy) Darwin's insights. Thus what follows concentrates on the Darwinian and developmental pillars and the interaction between them. This brings us back to the three all-important players: genes, individual organisms and populations.

How do novelties arise at each of these three levels? In conventional evolutionary theory, this question can be given a satisfactory general answer in only two of the three cases: genes and populations. So let's start with those. A gene is a stretch of DNA that acts as a code to make a protein – or part of a protein in some cases. The molecular details of this process have been worked out over the last half-century, ever since Watson and Crick provided the necessary foundation. Many unexpected complexities (like 'genes in bits', with the bits being separated by non-protein-coding DNA) have appeared. But none of these is essential for my story.

One important feature of genes is their almost complete constancy. Take the gene that makes part of the human haemoglobin molecule, for example (the 'beta chain', consisting of 146 amino acids). It has been known for many years that one particular mistake in just one of these amino acids causes the disease called sickle-cell anaemia. If this gene is in your family now, it has been there for countless generations in the past and will probably remain there for the foreseeable future. Not only that, but this gene exists in almost all of the 100 trillion cells of your body (albeit it is switched off in most of them), and in almost all of these it is the same. In someone with normal haemoglobin, the same principle applies: that normality is a feature of the past, the future, and almost all of the body's cells.

But this cannot be entirely true because we are all ultimately related. So at some fateful moment in history a normal haemoglobin beta-chain gene became a sickle-cell one. The fact that such changes occur, but very rarely, is why I described genes as having 'almost complete', as opposed to perfect, constancy.

Any change in a gene, whether its consequences are negligible, life-threatening or advantageous from the organism's point of view, is called a mutation. So here we have a single word of enormous importance. It is a cover term, an umbrella term, or whatever other phrase you would like to use, for all possible changes in all genes, and indeed in non-genic DNA too, though I'll sweep that aspect of the problem under the proverbial carpet. From this starting point you can create as many subdivisions as you like, and for various different reasons. Biologists will be familiar with spontaneous and induced mutations; with dominant and recessive mutations; with small ('point') and big mutations; and many other categories in addition to these few. If you are not a biologist, you may not have come across any of these – but, luckily, it doesn't matter. The important point is that we have a word that it is appropriate to use when we are talking about the origin of novelty at the level of the gene.

A gene, of course, is just a gene. A typical multicellular organism has thousands of them, and all are replicated as cells divide. And a population of such organisms adds another layer of replication. Let's look now at the question of how we describe the origin of novelty at the population level, before heading finally for our hardest challenge – the organism.

Because a population is necessarily a collective entity, while a gene is a singular one, there is a little more difficulty in describing the origin of novelty at this level. The question 'what is a new population?', when asked in an evolutionary context, can be answered in three ways. First, we can concentrate on how a novelty first appears in one individual in a population, regardless of its subsequent fate. Here, the answer is either the same as for genes (mutation), or alternatively 'immigration' in cases where the new variant gets into the population we are considering from another population. Second, we can ask how, in cases where the novelty is an advantageous one, it spreads through the population. Natural selection is the appropriate answer now, with perhaps a distant echo of 'genetic drift' somewhere in the background. This is the 'evolution as changing gene frequency' scenario. But finally, we can also ask how the population eventually becomes completely composed of the new variant at the expense of the old – something known in the trade as 'fixation' of the new variant. This needs no additional explanation. The same things that change a gene frequency from 45 to 46 per cent – natural selection and genetic drift – also change it from 99 to 100 per cent.

It seems a bit pointless to think of a novel population as one in which a new mutation has just occurred, both because in a large population the novelty is a miniscule thing and because it seems ridiculous to end up with the same term – mutation – to describe the appearance of novelty at both genic and population levels. It seems better to focus on changing gene frequencies, including the special and important case of fixation. If we take this view, and concentrate on adaptive changes rather than changes that don't affect adaptation, then the agency responsible for the introduction of novelty at the population level is standard Darwinian selection. As with mutation, there are lots of subcategories. Biologists are likely to have come across directional and frequency-dependent selection; positive and negative selection; visual and climatic selection, and so on. But again, as with mutation, the details are irrelevant here and if these are unfamiliar to you, it matters not.

* * * *

Simple questions, simple answers. What process causes the appearance of novelty at the genic level? Answer: mutation. What process causes the appearance of novelty at the whole-population level? Answer: natural selection. But what process causes the appearance of novelty at the organismic level?

At this point we need perhaps to remind ourselves that by 'organismic' we really mean 'developmental'. As noted earlier, there is no way that a new kind of creature can be made by changing one adult directly into another. The only way a new creature can be produced is by deflecting the course of development. So, what is the term for this?

Until recently, anyone faced with this question – say a student sitting an exam – could be forgiven for giving the answer 'heterochrony'. For the benefit of anyone who hasn't met this word before, it translates fairly easily – different timing. It refers to evolutionary changes where some process happening in the embryo – for example heart formation – gets shifted in time relative to other developmental processes, probably as the result of a mutation in one of the genes controlling the process concerned. Gould's *Ontogeny and Phylogeny* was very much centred on heterochrony;[2] he developed a 'clock model' to try to clarify its different subcategories. Moreover, whole books on heterochrony have appeared since, such as the one of exactly that title which appeared in 1991, written by Michael McKinney and Kenneth McNamara.[3]

When I first began to move sideways, academically speaking, into the new field of evo-devo, I was completely dissatisfied with this apparent answer. It seemed to me then, and the feeling has grown stronger with time, that it is just not possible to account for the origin of all evolutionary novelties at the organismic/developmental level – or even a majority of them – in terms of altered timing. And indeed it is often the most important changes that seem least explicable in this way. Some of the differences between human and chimpanzee morphology may be explicable in terms of heterochrony, as many have suggested. But in the grand scheme of things, these are minor differences, even if this seems an uncomfortable fact to some of us.

What about the appearance, in evolution, of the animal head? How do you explain the origin of a head merely through altered relative timing of developmental processes? Clearly, you can't. To be fair, most proponents of heterochrony would not argue that you could; also, much hinges on exactly how the term is defined. Nevertheless, I think that heterochrony's role has often been overstated. Of course, that doesn't mean to say that there are not many important evolutionary changes in development that involve an element of heterochrony combined with other things. But heterochrony alone is not enough. In fact, heterochrony is not an organism-level equivalent of genic mutation or population-level selection; rather it is more akin to a subcategory of one of these things, as encapsulated in the chapter heading 'It's not all heterochrony' in the book *The Shape of Life*, written by American biologist Rudolf Raff.[4]

* * * *

If heterochrony is just one subcategory of organism-level evolutionary change, what are the others? Well, where there's time, there's space. If things can be shifted earlier or later in the embryo, they can also be shifted from one place in the embryo to another. This is called heterotopy, and it has been recognized as a possibility for a long time, though strangely biologists have given it much less thought than its temporal equivalent. But these two do not exhaust the possibilities. In order to add the remaining two, let's focus on some particular microcosm of development in order to give the discussion some substance. Let's take a gene that gets switched on in a transverse 'stripe' across the body, such as one of the segmentation genes we looked at briefly in Chapter 4. In this context, heterochronic changes would involve the gene being switched on earlier or later, while heterotopic changes would involve its stripe being shifted a bit forward or backward along the anteroposterior axis, or changing position in other ways, such as becoming thicker or thinner or even becoming a patch rather than a stripe.

What else can happen? That is, supposing that there is no change in the timing or positioning of the stripe of gene expression, what other forms of evolutionary change in this developmental system are there? As far as I can see, there are just two other possibilities, and I have called them (in a paper published in 2000) heterometry and heterotypy.[5] These words were chosen both to be as self-explanatory as possible (different amount, different type respectively) and to link in as well as possible with the existing two terms. In relation to our 'stripe' gene, a heterometric change involves a greater or lesser amount of the gene's product being made, but in the same place and at the same time as before. A heterotypic change involves a change in the nature of the gene product (due to mutation, like all the other three categories), which might, for example, cause it to switch on or off a different range of downstream 'target' genes from before.

So now we have an exhaustive list, and, happily, it is only four items long. At the developmental level, novelties can be initiated in any of four ways: by altered timing, positioning, amount or type of gene product. And these four types of change also apply at other lev-els – for example altered cells rather than gene products. However, it is worth mentioning two complexities. First, there are many steps in the developmental process (or any part of it, like the formation of segments or limbs) between the switching on of the first genes involved and the realization of the final morphological product. It is entirely possible that (say) heterochrony at one step will lead to (say) heterotopy at another. That is, the nature of the change may alter as development proceeds. Second, even at one particular step, the change that occurs may in reality be a mixture of two or more of our four possibilities. Indeed, this may be the norm rather than the exception. However, such complexities should not deter us from per-sisting with a logical classification of the fundamental ways in which developmental change can come about in evolution.

* * * *

But if heterochrony, heterotopy, heterometry and heterotypy together form an exhaustive list of subcategories, what is it that they are

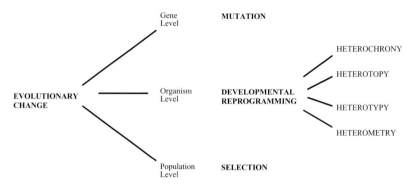

FIGURE 10 Logical relationships between developmental reprogramming and other types/levels of evolutionary change.

subcategories of? In other words, what is it that is the developmental equivalent of the umbrella terms mutation and selection that apply to the genic and population levels? Well, it's the easiest thing in the world to miss a single word in the vast scientific literature on a big subject like evolution. The number of words written about this subject since, say, 1859 doubtless runs well into the billions. So I may have missed it. But as far as I am aware, no one has coined such a word or phrase. If they have, it has made little impact and has passed quietly into obscurity.

This seemed to me an extraordinary omission from the scientific vocabulary. How on earth can we do equal justice to 'the organism' in evolutionary theory if we do not even have a cover term for all possible changes at this level, equivalent to mutation and selection at other levels? The short answer is that we can't. So we need a new term. When I first realized this a few years ago, I agonized long and hard over the best possible choice. In the end, I came up with 'developmental reprogramming'[5] (Figure 10). Not very inspirational, I'd be the first to agree, but appropriate nevertheless for the following reason.

Genes control the developmental process, but it also controls them. Indeed, the whole of development is a constant interplay between these two forms of control. There is, if you like to think of it this way, both a genetic programme and an epigenetic programme.

Also, the interplay between these two all-pervasive programmes can be influenced by a third factor (probably best not described as a 'programme'), namely the environment. This is sometimes only a slight influence, as when variation in food availability causes slight differences in body size between individuals. However, in a few cases it is a much more dramatic influence, as when an insect can end up either with or without wings solely because of the nature of the environment rather than any genetic change.

There is, though, one important difference between the genetic programme and both its epigenetic counterpart and those potentially important environmental influences. Only the genes are entirely inherited. This is what gives the necessary relationship between mutation and reprogramming. Development will only be reprogrammed in a heritable way if a gene has mutated. Of course, if 'housekeeping' genes with no developmental role mutate, development will remain the same. Even when a developmental gene mutates, the course of development will not necessarily be affected, because of the possibility of 'silent' mutations, where the change in the DNA sequence does not lead to a change in the corresponding protein (because of the so-called redundancy of the genetic code). Even if the protein changes, not all such changes will alter the developmental trajectory. To put all this succinctly, mutation is a necessary but not a sufficient cause of reprogramming.

It's time to revisit cascades. Recall that upstream genes control the activities of downstream genes. Let's think for a moment about a cascade of just two levels (and thus just one link). We'll call the upstream gene the controller, the downstream one the target. Let's focus on the target, and consider how the 'four heteros' can be produced. Heterotypy is the easiest to deal with. This will only occur in the target gene's product if the target gene itself mutates. But what of the other three? These could also occur through a mutation in the target, but are perhaps more likely to occur as a result of a mutation in the controller, given that the controller's product determines the switching on/off of the target. It is the interaction between the controller's product and the target's receptor sites for this product that matters.

*   *   *   *

Armed with our new terminology, we can now look afresh at the history of the organism's displacement from evolutionary theory. Back in the nineteenth century, most biologists were focusing on the organism. They were investigating developmental reprogramming, though of course they did not use this term. This was true of von Baer, who investigated comparative embryology but strangely not evolution (because he didn't believe in it). It was also true of Darwin, who investigated evolution (but not embryology at first hand, because he wasn't an embryologist). It was perhaps most true of Ernst Haeckel, who explicitly related the two processes – evolution and development – and worked primarily at the organismic level. And it was also true of Geoffroy Saint-Hilaire.

When theoretical population genetics began to be built in the early twentieth century, the organism was cast aside. The embryological component of evolutionary theory that, as we noted earlier, was far from absent in *The Origin of Species*, was entirely absent from the work of R. A. Fisher and many of his contemporaries. And although the organism began its comeback in 1977 with Gould's *Ontogeny and Phylogeny*, it did so in a rather incomplete way, with too much emphasis on heterochrony. That is not to say that I think I know what the relative importance of the 'four heteros' is – far from it. Neither I nor anyone else could claim such knowledge at present. But at least we can now pose the question. At least we have a certain minimum of terms to cover the actual biological processes that occur. We may yet need more, but terminology is best introduced sparingly. Let's get by for now on what we have: one umbrella term and four subcategories.

So, evolution works as follows: genes alter by mutation; development alters by reprogramming; populations alter through selection (and drift); new species arise when populations diverge to the point where they become reproductively isolated. Mutation, selection and reproductive isolation are already well represented in evolutionary theory. Reprogramming is not. We do indeed have a large, and rapidly accumulating, body of *data* on reprogramming, but it has not yet made the *conceptual* contribution that it can. We will now begin to examine the form this contribution may take, and, as will become clear, this will not end up being self-contained. Rather, it will force us to reassess how we interpret mutation and selection too. The quiet revolution is about to get louder.

8   Possible creatures

It's time to make another mental leap – from consideration of the actual world that we see around us, with all its actual creatures, to consideration of other possible worlds, with other creatures that might have been. I will argue that a 'possible creatures' approach radically alters our view of what determines the direction that evolution takes, and leads to a more inclusive understanding of evolutionary mechanisms generally. Although my approach of thinking about possible creatures has something in common with Stephen Jay Gould's famous metaphor[1] of 're-playing the tape of life', the emergent message will be very different. I do not deny the importance of historical contingency upon which Gould focused attention – indeed I will link up with it in due course. But my main message here is to do with interaction, not history.

Our starting point is that evolution is a two-step process: first, novelties appear – as a result of mutation and reprogramming; second, these novelties either do or do not spread through the population (and ultimately the whole species), depending on whether or not they are of increased fitness relative to their predecessor. This is not controversial; rather, all biologists are agreed that both of these steps must happen. The existence of variation – with the presence of two alternative types representing the minimum in this respect – is a prerequisite for evolutionary change. Without heritable variation, natural selection can achieve nothing. This is easy to see if you simply picture a population consisting of many individuals that are both genetically and developmentally identical. Whatever the environment throws at such a population, it will not evolve. It cannot adapt to any new conditions in which it finds itself, because the individuals are all the same. So, for example, if the climate warms up, the population will either stay the same or die. Since there are no variants that have a slightly better temperature tolerance than others, selection is impotent.

Luckily, few if any real populations are like this. Variation is the rule, not the exception. Most populations are in a continual state of flux because as the environment alters – which it rarely stops doing – some variants that were previously less fit become more so. Thus what was a decline in their relative frequency turns into a rise. If the environment keeps fluctuating this way and that, such evolutionary changes are likely to be minor and quasi-cyclical. But if conditions persist in some particular state – say elevated temperature – for a lengthy period, then evolutionary changes may accumulate into something more substantial, and that something will have a definite direction.

The prevailing view in 'mainstream' evolutionary theory is that this direction is not determined by the variation itself, but rather by the force – natural selection – acting upon it. If the average length of the legs of a certain type of mammal increases over a period of evolutionary time, this is not because the only variants that arose were those with longer legs. Instead the assumption is made, probably correctly, that variants arose in both directions, and that legs got longer because natural selection favoured those individuals with longer legs and acted against those whose legs were shorter. There are many possible reasons why such a form of selection might happen. One of the most obvious is that the creature we are considering is attacked by a predator, and longer legs give greater speed which in turn confers a slightly higher chance of surviving, reproducing and passing on the genes that tell the developmental process to make longer legs. Classic stuff; and fine as far as it goes. But now we have to go beyond it.

* * * *

I suppose there is a small chance that I am a distant relative of Alfred Russel Wallace; and also that although I am Irish and he was English our genes, or at least some of them, are/were Scottish. Wallace is a Scottish name, and, in a messy probabilistic way, genes and names tend to go together. Although in my case 'Wallace' is a Christian name, I was given it because my father's mother's maiden name was

Wallace. Scottish names are commonplace in Ulster as a result of the 'plantation' of settlers there several hundred years ago, causing the tensions that, even today, are still with us in the form of what are euphemistically called 'the Troubles'.

Anyhow, whether I am a distant relation or merely a fellow biologist displaced in time, I owe the great man an advance apology, because I am going to use some of his words to illustrate what I think is one of the main problems that has characterized evolutionary biology from his time to ours. This is the problem of regarding variation as omnipresent and essentially amorphous, which takes it 'out of the equation' in terms of determining evolutionary direction because, from the variation point of view, anything is possible. Here is what Wallace had to say on this issue in one of his essays[2] of 1870:

> Universal variability – small in amount but in every direction,
> ever fluctuating about a mean condition until made to advance in
> a given direction by 'selection', natural or artificial – is the simple
> basis for the indefinite modification of the forms of life.

It would be hard to find a clearer statement than this to the effect that selection is the sole determinant of evolutionary direction. Variation is a prerequisite, of course, but that's all; as long as the variation is there, selection will determine the direction in which it gets taken.

Let's now try to build a modern-day equivalent of Wallace's statement. It will not rival the elegance of his words, but it will help to give a more complete understanding of this general world view, which is still so prevalent today. A good starting point is human height. This is often used in genetics texts to illustrate the phenomenon of continuous (as opposed to discrete) variation. If we plot the frequency distribution of adult males or adult females for the human population of, say, Iceland, we will get the familiar bell-shaped or 'normal' curve with most values close to the average and fewer further away, as shown in Figure 11 (top). Admittedly, character distributions

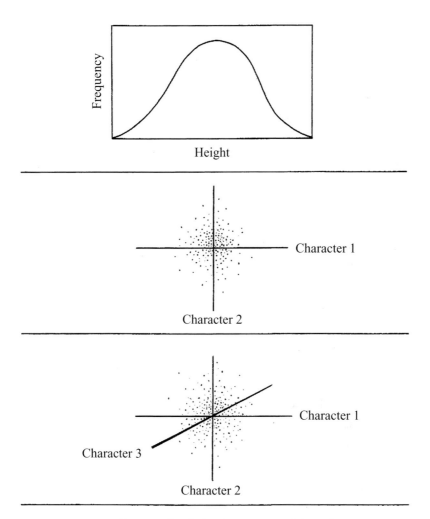

FIGURE 11 Patterns of variation in one, two and three characters.

for only adults don't make a very good developmental approach, but we can't do everything at once, so let's live with this restriction for now.

Neither humans nor any other creatures vary in only one character or dimension. Rather, multidimensional variation is the norm. But let's get there in stages. Moving up to two characters, the situation can be represented as shown in Figure 11 (centre). Since we now need both vertical and horizontal axes as the measurement scales for the two chosen characters, whatever they may be, frequencies need to be represented in some other way, and here I have chosen to represent them by the density of stippling.

This form of representation will work for any number of dimensions; it's just that the picture becomes harder to display within the confines of the two-dimensional page. Figure 11 (bottom) shows a three-dimensional version, with the third dimension projecting from front to back. A hologram would be better than a page for this, but the multimedia world has not yet impinged on the book world to quite that extent. Beyond three dimensions takes us into 'hyperspace' – not space at all, really, just a representation in which lots of axes representing lots of characters all project at different angles from a central cross-over point which is the average value of each. I won't even attempt to draw this – the head is a better place than the page to construct such multidimensional images.

If you can construct the appropriate image, what you should be seeing in your mind's eye is a shimmering cloud of points in hyperspace, approximately spherical and with the central region looking denser (darker?) than the periphery. This, then, is a sort of semi-quantification of what I believe Wallace had in mind when he wrote the words quoted above. It retains both of the important features that he emphasized: universality/all directions; and small/continuous. Or omnipresent and amorphous, as I described it earlier.

Of course, there are all sorts of complications both with this picture and with my description of it:

1. You can't reduce all the variation in a population or species to a single picture without causing a few problems.
2. Some characters vary in a discrete rather than continuous way. Some vary in both ways simultaneously, as in human eye colour where we can distinguish blue and brown, but also lots of shades and hues of each.
3. There is no developmental dimension, as I pointed out already.

4. The globular or spherical overall shape of the variation is dependent
on the scales you use for each character – if you want to stretch the
picture out in one particular dimension, a change in the scaling of
that character will do the trick.

And so on. But let's ignore all these things; they don't affect the
main point that I want to make – a point that is related to the one that
Swiss biologist Christian Klingenberg was making when he said[3] that
'the developmental system "channels" the phenotypic expression of
variation'.

However, I can't make that point quite yet, because there is one
more concept that we need to take on board first. This is the 'adaptive landscape', a form of imagery developed by that early American
population geneticist Sewall Wright, a contemporary of R. A. Fisher.
The adaptive landscape is an abstract concept and I don't want to
lurch straight from one abstract concept – variational hyperspace –
to another. Let's put something real in between as a sort of sanity-
retaining pause. And given where we're going, the best such thing to
put in between is the real landscape: hills, valleys, plains, etc. In a
word, topography.

* * * *

I grew up very close to the border between County Antrim –
Ireland's north-eastern corner – and County Down, which, appropriately enough, lies to the south. Although abutting, these two counties have strikingly different topographies. Most of County Antrim
is a flattish upland – the 'Antrim plateau' – with deep east–west valleys ('the glens') carved into it at various points by a series of rivers.
Most of County Down, in contrast, is fertile lowland, but with barely
a flat piece of land to be seen because it is composed of numerous
small round hills called drumlins for which apparently the retreat-
ing glaciers of the last ice age were somehow responsible, though the
details of exactly how glaciers produce drumlins have never been clear
to me. At the far southern end of County Down, the drumlins are
replaced by the altogether larger Mountains of Mourne, which are

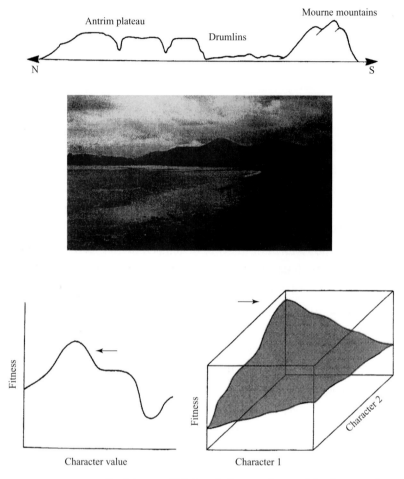

FIGURE 12 Real (top, middle) and abstract (bottom) landscapes. Top: Greatly oversimplified section, approximately N–S, of countries Antrim and Down. Middle: The Mountains of Mourne 'sweeping down to the sea' as an old song has it. Bottom: 2-D and 3-D adaptive landscapes. Arrows show maximum fitnesses.

steep-sided, craggy 'mountains' (just hills to an inhabitant of the Alps or Rockies) rising to almost 1000 metres (2796 feet, to be precise). Figure 12 (top) shows a cross-section through these two counties to illustrate their very different patterns of topography. This picture is, as you might expect, framed by the variables 'altitude' and 'distance'. All good solid stuff; nothing abstract here.

But Sewall Wright used just such a solid start to develop the abstract analogy that has been called, ever since, the adaptive landscape. This was a clever trick. Abstract concepts that have some link with the familiar stick in the mind much more readily than those that don't. And this one has stuck in a lot of minds for a lot of years.

To get from real to adaptive landscapes you need to do the following. First, you stop thinking about geography and start thinking about evolution. Second, you replace 'distance' on the horizontal axis with 'character value' (e.g. length of a bird's beak, thickness of a snail's shell). Third, you replace 'altitude' on the vertical axis with 'fitness'. Finally, and this last step is optional, you can make the whole thing three-dimensional by plotting fitness not against variation in a single character but rather over a two-character base plane (e.g. beak length and depth; shell thickness and diameter). Hypothetical examples of 2-D and 3-D adaptive landscapes are given in Figure 12 (bottom).

What do these new, abstract topographies tell us? Essentially, they tell us which character values, or which combinations of character values, are better than which. And by better, I mean fitter. In other words, the bearers of these values/combinations on average produce more surviving offspring than other variants. Thus evolution should move a population towards an adaptive peak.

Before proceeding further with the argument, let me mention a few options regarding the base of the landscape that I have so far neglected. First, it is possible to use either genetic or phenotypic variables to frame the base. I have opted for the latter. This is fine so long as we are dealing with *heritable* phenotypic variation; and I will assume that we are. Second, there is no developmental dimension, so really as character values change through ontogeny we should imagine a different adaptive landscape for each developmental stage. It's easy, even for a student of evo-devo, to fall into the old trap of thinking only of the fitnesses of adults, especially when there are already too many dimensions to cope with. And speaking of that, the base really

Character 1 value

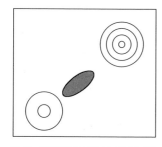

No evolutionary
change

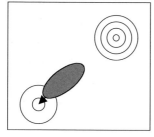

Evolution towards
bottom-left (lower)
fitness peak

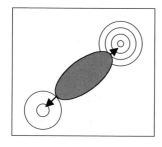

Evolution towards top-right
(higher) fitness peak (or
potential divergence
towards both)

FIGURE 13 The effect of the amount of variation (shaded) on the direction of evolutionary change. Circles are fitness contours, with innermost circles representing fitness peaks.

ought to be multidimensional, because organisms vary in so many ways, but life is hard enough already so let's not go beyond two.

* * * *

Now we need to put the structure of the variation and the structure of the landscape together. I first do this in Figure 13, where I show the influence of the amount of the variation on the course of evolution. In the example chosen, evolution does three different things despite the fact that the landscape is identical. The differences are entirely due to the different amounts of variation in the three cases. With little variation, the population stays put in the valley. Natural selection is indeed a 'blind' watchmaker. It cannot see mountains unless the population's variation is sufficient to overlap at least the beginnings of the foothills. With a bit more variation, the population climbs a smallish peak and stays there. With more variation still, the population can reach an entirely different peak, and might even split in two by 'attempting' to climb both peaks at once.

Let's now lengthen our timescale. Suppose that either of the bouts of hill-climbing that we have just witnessed corresponds to an evolutionary process taking 1000 years. What happens in the next thousand? Answer: it depends. Given (a) that the chunk of landscape we're looking at is surrounded by further landscape extending to the horizons, (b) that as the environment changes the peaks may shift (a seascape analogy has sometimes been used) and (c) that the amount of variation in the population may change, future evolution is rather unpredictable. But there is an important principle that can be clearly seen through all the complexities: the initial escalation of one peak or another may in the long term lead in completely different evolutionary directions. Yet the environment is the same. The only difference is in the amount of variation in the population. But this small difference may result in organisms of very different morphology in the distant evolutionary future.

* * * *

This, however, is a mere preamble. Personally, I'm far more interested in the effect of different *structures* of variation than in the effect of different amounts. And this is where it becomes necessary to criticize Alfred Russel Wallace. The mental picture that I have tried to engender, based on his idea of 'universal variation' is, as we have seen, a spherical or globular one, wherein there is a more or less equal amount of variation in all directions. Although I have referred to this as 'amorphous', perhaps this is not quite the right label. After all, a sphere has a shape, just as a dagger or a space shuttle has a shape. Perhaps a better term with which to describe that kind of globular variation is a 'null model'. This term, which is related to the statistician's 'null hypothesis', has been used by community ecologists to refer to situations in which certain kinds of structuring are absent, regardless of whether the 'null' state is entirely formless, which, in the nature of things, is unlikely.

Let's get practical again. Let's think about mammals, and about their leg lengths. Specifically, let's focus on the relationship between the length of the foreleg (or arm in our case) and the length of the hindleg. Now generally speaking there is a very strong positive correlation between these two measurements. This is easy to appreciate, and all that is required is a long list of mammal names and sufficient imagination to picture the creature concerned rather than just its name. The following list is intended to provide a wide coverage of the various mammal families, but avoiding those that have departed from the typical mammalian habitat of land and adapted their limbs for the purposes of flying (bats) or swimming (whales). So here goes: horse, dog, cat, mouse, elephant, giraffe, lemur, badger. Now picture a graph of average forelimb length plotted against average hindlimb length, with each of these, and many other, mammal species represented by a single point. Even if we use just our representative subset of mammals rather than an exhaustive list of all 5000-plus species, it is abundantly clear that our graph of forelimb against hindlimb length is going to give us a dramatic upward slope.

This is a statement of the obvious. A giraffe's forelimbs are much longer than their murine equivalents. Everyone knows this. And the same applies to hindlimbs. So if we were to plot a graph with only two points – one for giraffes and one for mice – we would get one point in the top right of our graph and one in the bottom left. Filling in the remaining types of mammal would hardly change the picture: the top right to bottom left picture would still prevail. Indeed it would be enhanced because many points would lie in between the giraffe and mouse extremes – see Figure 14 (top).

It may seem like a crazy question, but why do we get this pattern? The pan-selectionist view would be, no doubt, that only

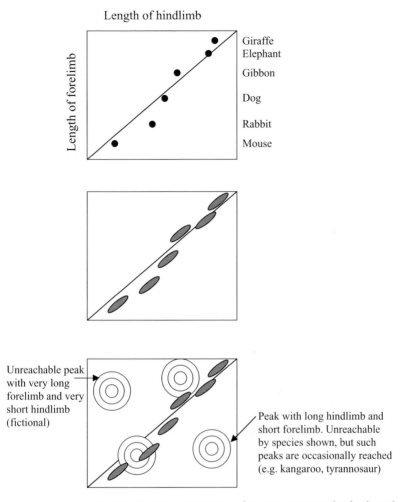

FIGURE 14 Top: Positive co-variation between average forelimb and hindlimb length among species. Middle: Within-species variation instead of just the average. Bottom: Adaptive peaks superimposed.

balanced designs are viable. Only when leg lengths are approximately the same in the front and back halves of the body do we get a reasonably 'fit' design, both in terms of engineering and in terms of natural selection. However, this is complete nonsense. Consider the following: rabbit (hindlegs about twice as long as fore); kangaroo (much greater asymmetry); and, though admittedly outside the mammalian realm, tyrannosaurs (hindlimbs an order of magnitude longer than forelimbs).

So, it is not necessarily true that unbalanced designs don't work. In that case, we need to seek an alternative reason for the predominance of balanced ones. And here is such an alternative, which has development at its core. The development of a limb is a highly complex process. Many genes are involved. The cascade of gene interactions that produces a forelimb is similar, but not identical, to the cascade that produces a hindlimb. Indeed, not everything is the same between the combined genetic, epigenetic and environmental 'package' that produces the left forelimb and that which produces the right. Otherwise, their lengths would be identical, which students of the horribly misnamed phenomenon of 'fluctuating asymmetry' (departure from perfect bilateral symmetry) know is very rarely the case.

At this stage in human knowledge of developmental processes, we cannot hope to dissect this complexity in its full horrors. However, one thing is clearly true, namely that some processes of development affect both forelimbs and hindlimbs. For example, in mammals, many hormones have a major effect on growth generally. And in some cases, more hormone equals more growth, regardless of whether the responding tissue is in the forelimb, the hindlimb, or the trunk. The story is about what Darwin called 'correlations of growth', not about natural selection.

This, in my view, is where Wallace went wrong. Indeed, he explicitly stated,[4] in 1897, that 'each part or organ varies to a considerable extent independently of other parts'. While no doubt both this view of independence and Darwin's view of correlation capture elements of the truth, correlations are not negated by the existence of some potential for independent variation; instead they are merely diluted somewhat.

So let's superimpose some intraspecific variation on the interspecific variation that we saw in the top panel of Figure 14. The new, combined picture is shown in the corresponding centre panel.

This superimposition immediately suggests a hypothesis as to why the latter (the interspecific variation) tends to fall along the upward diagonal: it falls there not because those are the only viable designs but rather because the nature of the developmental system structures the variation in such a way that this is the most likely result.

This point can be reinforced by bringing back the idea of adaptive landscapes. Let's suppose for a moment that adaptive peaks are scattered all over the place in the plane formed by forelimb and hindlimb lengths, as shown in Figure 14 (bottom). Which of these peaks can be reached on the basis of existing variation within any particular species? The answer is clearly that the fact that the variation is distorted – or *biased* – away from the 'null' globular pattern renders those peaks near the upward diagonal more 'reachable' than others in probabilistic terms. That is, *the structure of the variation determines, in conjunction with natural selection, the direction of evolutionary change. This directionality is not the sole preserve of selection.*

It's time for a couple of caveats. First, I'm not suggesting here or anywhere else that natural selection should be supplanted by developmental bias as the main agent of evolutionary directionality. Rather, I believe that it is their *interaction* that determines which way evolution proceeds. This is the interaction that I said at an early stage in this chapter was my main message, as opposed to Gouldian contingency. Second, we need to go back a step and ask why developmental systems tend to produce 'correlations of growth', rather than independence of characters, in the first place. Is this in some way inevitable, or is it in turn the outcome of selection in the past? A bit of both I'd guess, but who knows?

Finally, back to the 'long view'. We saw earlier that an initial small 'fork' caused by different amounts of variation could lead to very different morphologies in the distant future. Now we have seen that the *structure* of the variation is even more potent in this respect. We are not dealing with transient, short-term effects here. Rather, we are dealing with effects that will compound themselves over time. So the array of organisms that we see before us in the garden, the zoo or the rainforest may be just as much a product of developmental bias as of natural selection. This is a sobering thought.

* * * *

Let's now begin to connect a few things together; specifically, my emphasis on interaction, Gould's emphasis on contingency, and the idea of 'possible creatures'. Once, in a philosophical chat with a friend over a pint of beer, we ended up discussing a particular analogy for the flow of time. This took the form of a long narrow trough filled with water in which numerous different objects were floating around. The passage of time was illustrated, in this picture, by the trough freezing up from one end, as it would if placed inside a chamber with an end-to-end temperature gradient of fixed slope. If we start with all temperatures positive – say from 5 to 15 degrees Celsius, there is, of course, no ice to be seen. But when the temperatures are all lowered so that the gradient runs from minus 2 to plus 8, one end begins to freeze. As the temperatures are lowered further until they are all negative, the whole trough will freeze. As time proceeds in such an 'experiment', the interface between the ice sheet and the liquid moves gradually along the trough from left to right (or the other way round if your mental picture is the opposite of mine).

The difference between the past and the future is represented here by the difference between the 'fixedness' of all the previously floating objects in the frozen past and the fluidity of their counter-parts in the undetermined future. If we add some kind of swirling device, such as a series of powerful fans suspended over the trough, so that the not-yet-frozen-up objects are all swirling around each other in complicated ways, then this makes the difference between the past and the future more pronounced. It also helps us to picture the impor-tance of contingency.

At the time we call 'now', there is only one actual past. There are many other possible pasts that might have been but never happened.

There are many possible futures, and which out of these many will become tomorrow's frozen past depends on an almost infinite number of things. My 'possible creatures' notion refers to both the past and the future, despite this awesome difference between them. Indeed, they are the same thing looked at from different vantage points. The dinosaurs are for us part of the frozen past, but they were only one of many possible futures as far as a thinking trilobite from the Cambrian was concerned.

Let me now use this general way of picturing things to illustrate the importance of (a) developmental bias and its interaction with selection; and (b) Gould's historical contingency. Take an environment characterized by a particular adaptive landscape. Imagine that two separate populations of 'species X' are close enough that either may eventually colonize the area concerned. But imagine also that they have different structures of covariation of their various characters. The difference doesn't have to be a big one. It just needs to be enough to tip the balance between climbing one of two adjacent adaptive peaks or the other.

In this situation, the accident of which population migrates into and colonizes the area we are considering, with its own unique adaptive landscape, will determine the direction of evolution in the short term. It may also influence evolutionary directionality ever after the initial fork if the landscape is such that the initial choice of hill leads to divergent subsequent 'flow'. (In these adaptive landscapes, populations flow up, not down!) So, what determines the direction of evolutionary change, and the array of organisms that inhabit the actual world rather than some almost-actualized possible counterparts? The answer to this crucial question, in my view, is threefold: the structure of variation (developmental bias); the shape of the adaptive landscape (natural selection); and historical accidents (contingency). These three, and their interplay, determine the evolutionary future. Their relative importance will vary from place to place and from time to time. But notice that their interplay is not a conflict.

Bias and contingency can only help to choose one out of many possible 'advantageous' paths. Whales, as mammals that have returned to water in the actual world, have altered their forelegs back to fins. But perhaps there was another possible world in which the structure of the available developmental variation was such that they also went back from lungs to gills, and so had no need to keep coming up to the surface for air.

9 The beginnings of bias

There are two types of evolutionary change from the perspective of where the variation that is the starting point for all evolution 'comes from'. First, there is the type of change that I described in the previous chapter, in which the relevant variation already exists within the evolving population, in the form of what we call the 'standing variation'. This is exemplified by the variation that we observe between individuals within any human population in characters such as height, shape, strength, and so on. Second, there is the type of change that requires a new mutation to come along. An example of this is when newly conspicuous pale moths became vulnerable to predators against the blackened tree trunks of the industrial revolution, but had to await the appearance of a dark 'melanic' mutant moth before evolving towards a novel form of camouflage.

You might well want to question the wisdom of the 'first'/ 'second' order that I have just used. There is certainly a logic in reversing this order. When a population of any species first appears in any geographical region, it often does so by the immigration of a few individuals from elsewhere, followed by a period of rapid growth in the number of their progeny because of the relative lack of competition for resources in their new-found home. Such a population will have an unusually restricted amount of standing variation because the number of founding individuals was so low. Effectively, the population is 'inbred'. Its evolutionary potential may be quite small. For many potentially adaptive evolutionary changes to occur, it is 'waiting for mutation'.

This chapter deals with the influence of developmental bias in just such 'waiting for mutation' scenarios. The previous chapter dealt with the role of bias in the other kind of evolutionary change, based on 'standing variation' that was already present. Now we can see that, taking a historical view of the evolution of any natural population living in a particular place, it might have been more sensible to have switched the order of these two chapters. There is, however, a very good reason why I have proceeded in what you might call reverse-historic sequence. I should explain this before proceeding further.

In a sense this is a case of paying more attention to the history of biology than to the history of a population of moths or other creatures. Over the last century and a half of evolutionary biology, there has been a link between what people have thought is the important orienting force of evolution and whether they have thought that the most important evolutionary changes have had their origins in standing variation or in new (large-effect) mutations. Essentially, those who thought in terms of standing variation (e.g. Charles Darwin, Alfred Russel Wallace and R. A. Fisher) also favoured natural selection as the main, or even supreme, evolutionary driver. In contrast, those who thought in terms of new, large-effect mutations (e.g. William Bateson, Richard Goldschmidt) downplayed the role of selection to mere fine-tuning, and proposed a major role for mutation, and its effects on the developmental process, in steering evolution in particular directions.

Personally, I believe that this link is false. I think that developmental bias is an important determinant of the direction of evolutionary changes *regardless of* which sort of variation these changes are based on. Further, I believe that the 'micromutations' that are supposed to be the basis of the standing variation and the 'macromutations' proposed by Goldschmidt and others and not separated by any clear line of demarcation. Instead, there is a continuum of magnitudes of effect from one to the other, so that we can describe those mutations whose effects on the organism are of intermediate magnitude as 'mesomutations'. Finally, I imagine that most evolutionary changes are based on a combination of mutations of different magnitudes of effect, and that, as they proceed, they involve inputs from both the standing variation and newly occurring mutations.

Because I believe that the orienting role of developmental bias transcends the (false) micro/macro gulf, it made sense to introduce this orienting role in the context in which it is least expected. That way, I broaden the debate and make clear that I am not a 'saltationist' like Goldschmidt at the same time.

Just two final points before we proceed. First, when I refer to 'mutations' herein, I am not thinking as a geneticist but rather as a student of developmental reprogramming. Although a mutation is a change in the DNA sequence of a gene, what I am focusing on here is how such changes affect the organism. So, for example, 'big-effect' mutations are those that deflect the developmental trajectory more than 'small-effect' ones, but this says nothing of the magnitude of the change at the DNA level. Sometimes tiny molecular changes can have massive developmental effects and vice versa.

Second, I am referring throughout only to mutations occurring in those genes that help to control the developmental process – that is, those that are indeed capable of causing developmental reprogramming. There are plenty of genes that have no developmental role, and this is true regardless of whether we are talking about people, snails, insects, trees or whatever. Such genes are irrelevant to my story.

* * * *

Sometimes even populations with plenty of variation may find themselves 'waiting for mutation'. Although this sounds paradoxical, it is entirely possible and may in fact be quite a common situation. It can be pictured by using our old friend, the adaptive landscape. Suppose a population finds itself on what is, in adaptive terms, an extensive flat plain. There are hills just about visible in the distance; in fact the horizon is ringed with hills, ridges and mountains. Perhaps this is broadly equivalent to the real landscape that would be encountered by a small spacecraft that landed in the centre of one of the moon's major craters, such as Ptolemaeus.

It is not the absolute distance to the nearest hill that counts in this situation, measured, say, in centimetres of body length, or some 'hairiness' metric in relation to protection from the cold. Rather, it is this distance *relative to the extent of the standing variation*. If our population is a little blob whose outer extremities of variation come nowhere near even the beginning of the slightest slope, then it is, in a sense, 'stuck'. The standing variation in body size and all those other things is of no adaptive use. Fitter states exist but they cannot be reached. Inertia rules.

Unless the adaptive landscape changes markedly, the only thing that will permit adaptive evolution from such an unhelpful starting point is a mutation that takes the developmental system of its bearer far beyond the outer limits of the standing variation. In body-size terms, this would mean a dwarf or a giant. In general terms, it means an 'outlier', as statisticians call such things; that is, it is a creature that 'stands out from the crowd' in one or more of its characters, even though the crowd is by no means a clone.

There's a can of worms here that I'd prefer not to open; but let's at least read the writing on the outside that describes its contents. On several occasions in the history of evolutionary biology, as noted above, prominent figures have argued that the standing variation – the small continuous stuff of the normal curve – is irrelevant to grand-scale evolution and need only be taken into account by those concerned with races, varieties, or other forms of minor intraspecific differentiation. The geneticist Richard Goldschmidt took this view. His *Material Basis of Evolution*,[1] which is the most complete statement of his theories based on 'systemic mutations' and 'hopeful monsters', was published in 1940.

As the modern synthesis was put together, and as it 'hardened' over time, Goldschmidt became everyone's *bête noir*. He is partly to blame for this because he did come up with an extreme theory – though some of his later publications suggest that it became less extreme with age. But the pro-synthesis folk, in their enthusiasm to gang up on Goldschmidt and to de-bunk his idea that most important evolutionary changes were based on mutations with individually enormous effects, became too extreme in the opposite direction. Their obsession with the all-importance of tiny-effect mutations and their denial of any significant role for 'macromutations' in causing sudden evolutionary leaps, or 'saltations', has characterized mainstream evolutionary thinking from the 1930s to the present day. A recent example is the claim by Richard Dawkins[2] that there are 'very good reasons for rejecting all such saltationist theories of evolution'.

That's the end of the writing on the label; I'm now going to proceed without opening the can. The reason I can do this is as follows. There are some mutations whose effects are both huge and bizarre. An example is the one called 'antennapedia' that gives flies an extra pair of legs. These legs, as the mutation's name suggests, grow out of the head where normally the antennae would grow. This mutation has not contributed to evolution. We know this partly because it makes flies very unfit and partly because there are no groups of flies (either extant or extinct) that have legs growing out of their heads. So, I will ignore such gross changes. I will focus instead at some intermediate magnitude – on mutations that make their bearers stand out from the crowd, but not by too much.

Now, I want to tell you what I and a few others have begun to think about evolutionary changes with this kind of basis. But, as in the case of 'standing variation', it will make more sense if I put it into a historical context. This time, Wallace's work in the nineteenth century is not the right context. He and his contemporaries knew a lot about continuous variation, but they knew little of mutation (though they were aware of the existence of the odd phenotypes that they called 'sports'). To get the correct historical context for the 'discrete' case, we need to move ahead about half a century to the birth of population genetics and in particular to the work of R. A. Fisher.

Fisher was an out-and-out theoretician. His main achievement, as I mentioned earlier, was the 'mathematicizing' of natural selection. His results took the form of equations. Fisher's Oxford-based protégé, E. B. ('Henry') Ford, who wrote *Ecological Genetics* (several editions and many printings starting from 1964 and covering a couple of decades[3]), was, in contrast, a field worker who preferred butterflies

to equations. In particular, he worked on variation in the pigmentation patterns on butterfly wings. Despite their different approaches,
Fisher and Ford were of one mind when it came to the cause (in the
singular, notice) of evolutionary directionality. A couple of quotes will
help to illustrate this point.

In 1930, Fisher[4] spoke of a 'logical case for rejecting the assumption that the direction of evolutionary change is governed by the direction in which mutations are taking place, and thereby rejecting the
whole group of theories in which this assumption is implicit'. A few
decades later, Ford[3] confidently stated that 'if ever it could have been
thought that mutation is important in the control of evolution, it
is impossible to think so now'. And in case we were in any doubt
about what agency was responsible for determining the direction of
evolutionary change, Ford goes on to say that 'living organisms are
the product of evolution *controlled* not by mutation but by powerful
selection'. (The emphasis is his, not mine.)

* * * *

To my mind, the problem with the Fisher–Ford approach is that it
envisages a conflict between mutation and selection. It sees a need to
choose one or the other as the factor that determines the direction in
which evolution proceeds. This was probably, in each of their cases, an
understandable reaction against extreme mutationist views in which
'micromutations' were deemed irrelevant or even uninheritable, and
Darwinian selection was a mere epiphenomenon in the history of
life – something that just fine-tuned the products of the huge macromutational leaps that were the stuff of major evolutionary change. So
Fisher was reacting against de Vries & co. while Ford was reacting
against Goldschmidt. They both took an almost anti-mutation view.
Both acknowledged, as they had to, that mutations were the ultimate
source of variation; but having said that, both dismissed mutation as
having any other role, especially a directional or 'controlling' one.

There is, however, no conflict. The view that I proposed earlier
for the 'evolution-by-standing-variation' scenario is equally applicable
here. It just takes a slightly different form. In the previous scenario,

evolutionary direction was determined by the structure of the variation, natural selection, and the interplay between the two. In the present case, I will make an exactly parallel argument: that evolutionary direction is determined by the 'structure' of mutations, natural selection, and the interplay between the two. What I mean by the 'structure' of mutations will shortly become apparent.

This is, again, a case of considering many possible worlds rather than only the actual world. To do this, I'd like to go back to the lunar crater analogy. Suppose our lunar lander (or population) is sitting in the middle of a flat crater base. Suppose that the crater is not one of the biggest ones, but rather one of intermediate size, so that we can avoid the Goldschmidtian can of worms and think in terms of biggish mesomutations. Now suppose that the crater's rim is a rather jagged one that rises to almost mountainous proportions at some points, while other parts of the rim are just a gently undulating ridge, as illustrated in Figure 15. Now forget the lander and the crater and think in terms of a population and fitness plain/peaks. We are now in that classic 'waiting for mutation' pause. The population eats, sleeps, reproduces, experiences mortality, and so on. That is, it does all the normal ecological things that populations do. But in evolutionary terms it does nothing, with the possible exception of a few irrelevant random meanderings as a consequence of genetic drift. It may continue to do nothing for thousands, maybe even millions, of years.

Then, suddenly, along comes a mutation. To begin with, it has to run the gauntlet of surviving its initial period of very low frequency when it is at great risk of being lost despite its conferring increased fitness – after all, mutations usually occur singly and a single individual is not a reliable vehicle. The first grass-green beetle in a population of shiny black ones may disappear under the proverbial cow's hoof, taking its potential advantage of predator-deterring camouflage to an early grave, and leaving its fellow beetles to wait another millennium for a second chance of becoming invisible to birds.

But let's be optimistic. Let's suppose that our mutation survives this fragile early stage and increases in frequency. It can do this rapidly

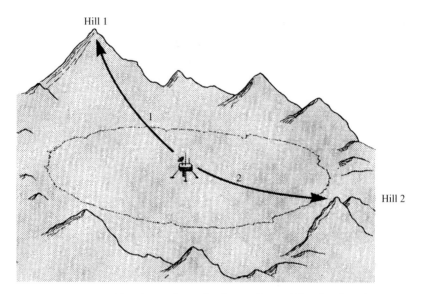

FIGURE 15 Population/lunar lander in the middle of a flat plain surrounded by adaptive/real peaks. Depending on which of mutations 1 or 2 occurs first, the population will end up on hill 1 or hill 2.

because it confers a large increase in fitness. So the time to fixation is short. Of course I have assumed that new mutant individuals are reproductively compatible with existing 'non-mutant' ones, but that's not an unrealistic assumption: many bizarrely mutant fruitflies can still reproduce with their non-mutant counterparts. I have also assumed that the floor of the crater is flat – in other words that there are no deep fitness troughs into which the offspring of a wild-type/mutant mating might fall as the population is *en route* to the hills. Again, this is not too unreasonable an assumption: adaptive landscapes come in all shapes and sizes, and some of them will be like this, even though others will have troughs as well as peaks.

It should be crystal clear that in a situation of this kind the direction of evolutionary change is determined by mutation. If a mutation occurs in direction 1 in Figure 15, the population ends up on hill 1. If, alternatively, a mutation occurs in direction 2, the population will end up on hill 2, at the opposite side of the crater and representing a very

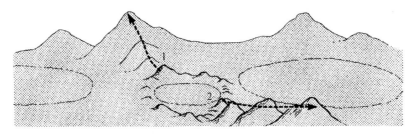

FIGURE 16 Possible long-term divergent evolutionary directions taken as a result of the initial shift towards hill 1 or hill 2.

different change in the developmental system. Now if we pan out and look at the landscape beyond the confines of our crater (Figure 16), we can readily see that there is a good chance that a population that starts at hill 1 will head off, in its subsequent evolution, in a very different direction from a population that starts at hill 2. *These evolutionary routes may diverge forever.* Not quite the picture that Fisher or Ford had in mind.

*   *   *   *

But I am falling into a familiar trap. Mutation may *look* as if it is solely responsible for directionality here, but it is not. As before, three factors are involved and they are the same three: mutation, selection and contingency. The various peaks around the edge of the crater represent various possible futures. They are possible in the sense that they represent states of enhanced fitness, so selection will happily take our population to any of them if the right mutation, delivering the right reprogramming, comes along. Mutations leading to troughs, if there are any, will perish. And mutations leading a short way across the plain will probably perish too, though by chance a few will survive. So the hills of the adaptive landscape give rise to a 'possible creatures' subset of the set of all combined character states, while the occurrence of a mutation and reprogramming in a particular direction selects the actual creatures from that subset. In other words, mutation/reprogramming and selection are working *together*. But what of contingency? And what of my obscure phrase 'the structure of mutation'?

These are in a sense the same thing. Let me explain. It is often said that mutation is 'random'. As a generalization, this is nonsense. It is a phrase that has been used casually beyond the domain in which it first arose and in which it is entirely appropriate. This is the domain of the relationship between the direction in which mutations occur (in terms of their effects on the developmental system) and the directions of change that would lead to enhanced fitness. I have anguished over many a first-year exam script where the student author proclaims that melanic moths first arose after the industrial revolution turned the trees black; somehow, mutation 'knew' that this was now the right way to go. Equally, I have smiled over the correct answer found in other scripts – that melanic moths have turned up by mutation every so often since the first moth appeared aeons ago, and that most were conspicuous and disappeared into the mouths of predators taking their mutant genes with them. But one turned up in altered circumstances, was cryptic and overlooked by predators, and the rest, as they say, is history – except that the nature of the selection is beginning to look a bit more complex than it did at first.

But there are many other ways in which mutations are *not* random. One of these, in particular, is relevant to our current story. Take the two mutations that led to hill 1 and hill 2. I am focusing, herein, on mutations that affect the developmental system – that is, that cause developmental reprogramming. In the case of these two specific mutations, the reprogramming is clearly major, but of fundamentally different kinds. In both cases the altered gene product – probably a protein that switches 'target' genes on or off – exerts its effects on the ontogenetic process in a complex way, with many knock-on effects further downstream in the developmental cascade. Ultimately, the adult has a different morphology, though of course the morphology of those many intermediate developmental stages will be different too; as ever, we need to keep in mind that selection does not act only (or even mostly?) on adults.

Given the complexity of developmental reprogramming – and this phenomenon is still largely a 'black box' at our current state of knowledge – it is possible, even likely, that some forms of reprogramming can be produced by many mutations, while others are produced by only a few. That is, some changes are 'easier' to achieve than others. So even if the rates of occurrence of different mutations at the molecular level are equal (and usually they won't be), the result will be a higher probability of producing some mutant individuals than others.

Since all probabilities of mutation are very low – generally below one in a million per generation – the way that a difference in the probability of mutationally induced reprogramming will manifest itself is in the frequency, and the order, of occurrence of the different mutant forms. This is crucial, because given the long periods in between one such mutation and the next, if a 'mutation 1' leading to hill 1 occurs first, the population may go to fixation in that direction long before a 'mutation 2', that would have led to hill 2, occurs. By that time, our population may have wandered off into some other part of the landscape from which point hill 2 is no longer an attractive (i.e. fitter) option. It has disappeared from the subset of possible worlds.

This is where contingency comes in. If one form of reprogramming is 'easier' than another, it will tend to occur before one producing the alternative form of reprogramming. So evolution will tend to go in that direction. But this is indeed just a 'tendency'. The whole process is a probabilistic one. With billions of populations evolving in billions of environments, there will inevitably be cases in which the less probable happens first. A higher probability does not determine a precise order of occurrence. So, as with all real scenarios in the real world, contingency may step in and send evolution in a direction that would not have been predicted.

*   *   *   *

That's about it, apart from a few niceties. The phrase 'developmental constraint' is often used in an evolutionary context, as I noted in Chapter 2. I have come to hate this phrase. In most authors' hands it implies a negative role for development and a positive one for selection. The picture I have painted above is hardly compatible with such a view.

This is why I recently coined the term 'developmental drive', which emphasizes the positive role that mutationally induced developmental reprogramming can have on the course that evolution takes.[5] Both of these – constraint and drive – are subsumed under the more general heading of developmental bias. And the related term 'mutation bias' has been coined recently by the American biologists Lev Yampolsky and Arlin Stoltzfus, to whom I referred earlier.[6]

So we end this chapter where we began the previous one, with possible creatures. The Fisher-and-Ford view was too limited. It focused only on actual creatures. That narrow focus led them, and many others, to overstate the role of selection and to think of it as the only agency that is responsible for our planet being inhabited by the creatures that are indeed here, as opposed to those that might have evolved but didn't – like whales with gills. Anyone who chooses to accept my broader view of 'possible creatures' will come to a very different conclusion: that developmental bias, natural selection and historical contingency are all inextricably intertwined; and that together they compose a compound causal agent that has determined the course of evolution for the last billion years, and will continue to do so for the foreseeable future. Perhaps a similar form of causality also operates in other biospheres,[7] if there are any 'out there', beyond the confines of our own solar system.

10 A deceptively simple question

Following the mind games of the last two chapters, we are now as familiar with adaptive landscapes as with real ones. But, taking a step backwards, what, in the first place, is this thing called 'adaptation'? There is a huge literature on this, both scientific and philosophical. Most of it I will pass by as it is not central to my task herein. But there is one particular issue that we now need to confront: how inclusive is 'adaptation'? More specifically, does 'adaptation' include both external and internal aspects? Does it include both the 'fit' between a bird's beak and its food supply (external) and the 'fit' between one of its bones and another interconnecting one (internal)? At the molecular level, does it include both the enzymes that digest a horse's food (external, because the digestive tract is a tube of the external environment projecting through an animal from mouth to anus) and the interplay between a signalling molecule and its receptor deep within the horse embryo (internal)?

The external aspect of adaptation is, clearly, adaptation to the environment; so it is often referred to as ecological adaptation. The internal aspect is adaptation of one body part to another; so it is often referred to as coadaptation. But unfortunately, the casual use of 'adaptation', naked and unqualified, is common in the biological literature. When authors use this word in an unqualified way, what do they mean? Specifically, are they using it as shorthand for ecological adaptation, or are they using it in a broader way, to include coadaptation too?

Regrettably, the answer depends on the author. Different authors use 'adaptation' in different ways. I suspect that there are even some authors who use the term in different ways on different occasions, either because they are blissfully ignorant of the potential

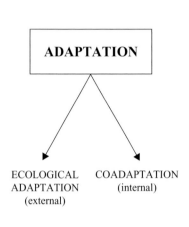
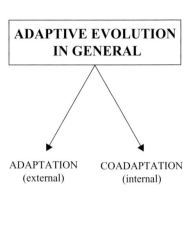

FIGURE 17 Alternative usages of the term 'adaptation', and the logical thought structures with which each is associated.

confusion, or else because they are assuming that the correct choice of meaning will be obvious from the context, which is a rather unwise assumption.

I illustrate these two usages, and the different logical relations that they imply, in Figure 17. Clearly, the relations depicted on the left and right sides of the figure are fundamentally different, and ought not to be confused. In my view, fuzzy thinking and inconsistent usage of 'adaptation' have caused major problems. Personally, I will opt for 'adaptation' as an overall cover term. I will use either 'external' or 'ecological' to qualify the term when used in one way; and I will use 'co' as a qualifier when using it in the other. In other words, I will stick to the logical structure implicit in the left-hand diagram in Figure 17.

Terminology in itself is boring. But this is not a sterile exercise in pedantry – far from it. As I hope to show, the confusion over the meaning of 'adaptation' has had far-reaching consequences for the nature of evolutionary theory. It has, in my opinion, been one of the factors responsible for the 'squeezing out of the organism' that I am trying to reverse in this book; because the original, broad 'adaptation', as used by Darwin, was replaced, by Fisher and others, with the narrow, 'external-only' version. A couple of quotes will illustrate this point.

First, Darwin,[1] in *The Origin of Species*: 'How have all those exquisite adaptations of one part of the organisation to another part, and to the conditions of life, and of one distinct organic being to another being, been perfected?'

Good old Darwin! Not only does he include the internal – co-adaptational – side of things, he even puts it first. And just for good measure, when it comes to the external aspect, he leaves us in no doubt that he means to include adaptation to both the abiotic (e.g. climate) and the biotic (e.g. predation) environment. Crystal clear in his inclusivity.

Now, Fisher, some seventy years later,[2] in *The Genetical Theory of Natural Selection*: 'An organism is regarded as adapted to a particular situation, or to the totality of situations which constitute its environment'. Notice that Darwin's inclusive view has gone. In its place, we have the narrower use of 'adaptation' to refer only to the relationship between the organism and its external environment. Fisher wasn't just a pan-selectionist, he was a pan-externalist. And his influence was such that the modern synthesis, as it developed, was imbued with this externalist emphasis.

* * * *

As ever, a school of thought is heterogeneous, and not all of its adherents take exactly the same view of things. Some neo-Darwinians *have* discussed coadaptation. However, (a) these discussions have usually occupied a minuscule space compared with that devoted to external adaptation; and (b) they have tended to focus more on coadaptation of genes and gene complexes than on coadaptation of organismic structures. Adding in genes to the picture is fine, of course; but squeezing out the organism is not.

This is another aspect of the lop-sidedness of 'mainstream' evolutionary theory that I have referred to before and that I aim to remedy. So from here on I will focus not on external adaptation, but rather on coadaptation. Anyhow, evolutionary texts are full of externalist stories – melanic moths, Darwin's finches, cryptic and not-so-cryptic landsnails, fly larvae with enzymes adapted to detoxifying the alcohols that abound in their rotting-apple home, and so on. So if you want to read about these you have plenty of choice of books. Books on coadaptation, in contrast, are rather hard to find.

Now don't get me wrong here. I have nothing against external adaptation. Not only did I start my academic life as an ecologist, but, as I moved from there to population genetics (and before moving on again to evo-devo), adaptation to the environment was my focus of attention. Not only that, but, as a Ph.D. student, I personally worked on those not-so-cryptic landsnails, and was surrounded by people working on other case studies of ecological adaptation. This is all good stuff, and evolutionary theory needs it. But what it doesn't need is yet another author giving yet another 'accessible' account of these stories. What it needs much more is someone to argue for what has become the underdog – internal organismic coadaptation. Hence the following.

So, where do I start? Well, let's begin by having some fun. Let's attack a pan-externalist 'strawman'. Now for those of you who are unfamiliar with this particular type of beast, a strawman is usually a cause for concern. An author who argues against 'theory X' by attacking a stupidly oversimplified version of it is often said to be attacking a strawman as opposed to the real thing, and is rightly criticized for doing so. (I suppose in this politically correct era I should use 'strawperson', but I just can't bring myself to do it. Sorry.)

In this case, the choice of a strawman target is not a substitute for attacking the real thing – when necessary I'll attack that too. But it allows us an easy route into a difficult topic; so it is a temporary means to what I hope will become – in terms of its effect on evolutionary theory – a permanent end.

Picture a conversation with a bespectacled pan-externalist strawman, who is sitting in a straw armchair listening attentively to what you and I have to say. First, we bring natural selection into the conversation. The strawman beams. Next, we progress to adaptation. The beam remains. Then, we narrow down to coadaptation. At this point, the strawman's beam becomes less pronounced, and we can perhaps detect a little apprehension about where the conversation might go next. Then, in connection with the evolutionary forces producing coadaptation, we use the phrase 'internal selection'. Now, the strawman's expression turns to annoyance, and he goes a deep purple. We then, foolishly, make the bold statement that 'internal selection, and indeed other internal factors too (like developmental bias), make a significant contribution to evolutionary direction'. We are only saved from being chased out of the room by the angry strawman by the fact that he undergoes spontaneous combustion.

So, a fun story; but the point am I trying to get across here is this. There is a continuum all the way from external adaptation (accepted by almost everyone) through coadaptation to internal selection to other mechanistic internal processes like developmental bias to mystical internal 'urges' (old-style orthogenesis; accepted by virtually no one). As we move along this continuum, we go from the orthodox to the heretical. However, in my view, all but the last – the 'mystical internal urges' that we never got a chance to discuss with our strawman – deserve to be in the realm of the orthodox. But two of them, namely internal selection and developmental bias, are not. I believe that the reason why many neo-Darwinians, both real and straw, are hesitant to take these on board is that they feel that these concepts are tainted by association. That is, because they are adjacent to the mystical urges, they are perilously close to falling out of the scientific realm altogether. Well, in the last two chapters I made the case for developmental bias; here I make the case for internal selection.

* * * *

The 'inventor' and champion of internal selection was the English writer Lancelot Law Whyte. The fact that he was a philosophical polymath and not specifically a biologist probably acted against his acceptance by the biological community of the time (the 1960s). He wrote several papers and one book[3] – *Internal Factors in Evolution* – on this subject. At first, he called it 'developmental selection'; but by the time the book appeared he had switched to calling it 'internal selection'. This was perhaps a mistake – we'll see.

I'll try to give you the gist of what Whyte was saying, without clouding the issue by also explaining the points that I think he got wrong. His book is a curious mixture; I found myself strongly agreeing with most of it but strongly disagreeing with a few specifics. Here I will focus on the former.

The key question is: what is the cause of fitness differences among the individuals in a population? It doesn't matter what the population is – pondsnails in a pond or flies in a forest – the question is an entirely general one. It has two possible answers. First, it may be that the fitness differences are due to relations with the external environment. Take the flies in the forest, for example. Suppose that the forest has shifted to a higher ambient temperature due to global warming (seemingly a myth as I sit here writing this in mid-July with the outside temperature struggling to get above 10 degrees Celsius, 50 degrees Fahrenheit). Suppose also that there is considerable variation in body size among the different flies of which the population is composed. Flies, like many other small terrestrial invertebrates, fight a continual battle with desiccation. The warmer it gets, the faster they lose water, and the worse the problem becomes. But, as I learned when a first-year undergraduate (or possibly even earlier): the bigger you are, the smaller your surface area is in relation to your volume. Since water loss occurs from the body's surface, if you inhabit a hot, dry environment, it is better, other things being equal, to be bigger. So in our hypothetical fly population, selection favours larger flies, and the average body size of the population will tend to increase, as long as it is unopposed by other factors such as insectivorous birds preferring bigger prey.

In this situation, the fitness differences among the members of the population are clearly produced by the environment. This, then, could be called 'external selection', and what it produces is one particular form of ecological adaptation.

The second possible answer to the general question of what produces fitness differences is the working relationship between different parts of the organism. Let's stick with the flies for an example. Suppose that, as well as varying in body size, the flies also vary in the way in which their wings are connected to their thorax. Nothing major, just the usual stuff of standing variation – very slight quantitative differences in the morphology of the wing/thorax joint.

Now suppose, and it seems entirely reasonable to do so, that this variation in the wing/thorax joint affects, albeit slightly, the flies' ability to fly. This could take the form of affecting the rate of the wing beat or a fly's manoeuvrability in situations where flying round tight corners is required. In this situation, the population will evolve towards better-integrated joints. But the selection that is involved in this evolutionary change is unrelated to the warming up of the forest. Indeed it is unrelated to the forest itself. It is one of those self-evident facts of life that flies fly. They do so for all sorts of reasons, all linked with survival. Whether flight is used to find food or a mate, or to escape a predator, slight differences in flight ability will affect the probability of success or failure. If our fly population migrates from its forest to a neighbouring meadow, flight is still required. So although flight takes place in 'the environment' in a general sense, good flight ability is not just an advantage in any one particular forest or meadow, it is an advantage in all of them. The exact degree of advantage might, perhaps, vary a bit from one environment to another, but we'll ignore that for the moment.

So in this situation, the fitness differences between individuals are not caused by any specific feature of the environment. Rather, they are environment-independent, or, more accurately, quasi-environment-independent. The selection that is taking place here is

Whyte's 'internal selection'. The name he ended up giving it is perhaps misleading, because it could be interpreted as selection going on within the organism. But of course the correct interpretation is selection going on in the environment (as usual) but due to fitness differences that are caused by different degrees of internal integration of the organism. The Connecticut-based biologists Guenter Wagner and Kurt Schwenk, who, like myself, have been impressed by Whyte's work, put it as follows:[4] 'internal selection travels with the organism wherever it goes'.

Personally, I see nothing really controversial here; yet when I have used the phrase 'internal selection' in conversations with neo-Darwinian colleagues I have often met with a raised eyebrow. And when consulting fellow 'evodevologists' about having conversations about internal selection with neo-Darwinians, I have often encountered the phrase 'red flag': that is, they have warned me that I may upset the neo-Darwinian bull.

I think that the lack of incorporation of Whyte's important work into mainstream evolutionary theory may lie in his choice of phrase. I understand his rationale for broadening out from his initial use of 'developmental selection'. After all, some internal selection, including our fly wing example, concern the integration of adult structures rather than developmental processes, though of course the former are the end results of the latter. But I think that Whyte overlooked the fact that this shift to 'internal' selection carried with it a risk of misinterpretation and resultant rejection.

* * * *

This is where we need to expand on the point made above, namely that in cases of what appears to be internal selection, the degree of fitness advantage of one variant over another may not be completely constant regardless of the prevailing environmental conditions, but rather may vary slightly from one set of conditions to another. It doesn't take a genius to spot the fact that this small and entirely reasonable admission gets us into a whole lot of trouble. Well, temporary trouble, anyhow; think of it as a river of trouble that we must wade through in order to get to the more enlightened understanding that lies on the other side.

The problem is this. There is probably no such thing as pure external selection or pure internal selection. If these phenomena exist at all, it is as opposite ends of a continuous spectrum characterized by a gradually shifting ratio of the importance of (a) specific features of the environment and (b) environment-independent features of organismic integration (both developmental and functional) in determining the fitness differences among the organisms within the population. If the effects of the environment are overwhelming, the situation may for convenience be labelled 'external selection'; in the converse case, 'internal selection' may be used, again as a label of convenience. But the truth is that there is a continuous range of types of selection from one extreme to the other.

In an earlier book, *The Origin of Animal Body Plans*,[5] I developed a way of picturing this range, in the form of something called, accurately but rather clumsily, the 'trans-environment fitness profile'. Some examples of such profiles are shown in Figure 18. Let me explain.

Let's pick an arbitrary number of environments in which a population of flies, snails, eels, owls or dinosaurs might find themselves. Say eight. That's unrealistically low, but it doesn't matter. Let's suppose that these eight environments are all characterized by different ecological conditions. We don't need to bother yet about the nature of these differences. To simplify matters, let's also make the unrealistic assumption that the population only has two variants – an 'old' one that has been around for some time, and a 'new' one that has arisen recently by mutation and, although the mutation confers some benefit and is thus favoured by selection, is still only present at relatively low frequency.

The top picture of Figure 18 shows pure 'external selection'. An example would be the case of a new variant whose external pigmentation pattern is so well camouflaged as to make it almost invisible to a particular predator that, as it happens, is only found in

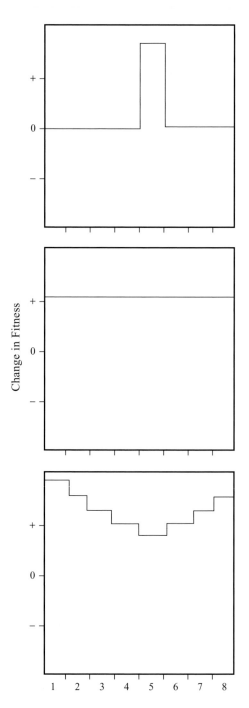

FIGURE 18 Possible trans-environment fitness profiles of a new variant. Top: Pure external selection. Middle: Pure internal selection. Bottom: Mixed-mode selection. 1 to 8 represent eight different types of environment. +, fitness increase; 0, neutral; –, fitness decrease.

Change in Fitness

'environment 5'. So that variant has a massive fitness advantage over the other in environment 5, but elsewhere it has no advantage at all and thus is, to all intents and purposes, selectively neutral.

The second picture shows the opposite: pure 'internal' selection. Here, the new variant is characterized by such an improvement in its internal integration that it has significantly elevated coadaptation across the board – that is, in whichever of environments 1 to 8 it finds itself.

The bottom picture, however, is more realistic than either of those above it. Here, we find ourselves somewhere in the middle of the continuum. There is an element of internal and external causation of the fitness differences, so the profile is neither a sharp peak nor a flat plain; rather, it fluctuates up and down a bit as the population moves from one environment to another. My guess is that most real selectively driven evolutionary changes are like this. In a way, that's a statement of the obvious, because if you accept the idea of a continuum – and it's hard not to – then the 'pure' cases at the ends represent a tiny proportion of all the things that could happen. So on straight probabilistic grounds, 'mixed-mode' selection should be the rule rather than the exception.

11 Development's twin arrows

It's now time to bring development back into the picture in a more explicit way. As we noted in the previous chapter, internal selection can occur based on fitness differences between adults, so there is no logical necessity for this process to be linked to development other than for the obvious reason that adults arise through the developmental process. But there is another angle on all this, which does indeed suggest that internal selection and development may have a particularly close relationship.

By the way, you will notice that I have debunked both 'internal' and 'external' selection in their pure forms, yet here I am starting to use these terms again as if I don't believe my own argument for a continuum. I must therefore stress that I *do* believe my argument, but I also need to be able to write in a relatively straightforward, non-cumbersome way. So phrases like 'selection in which 75 per cent of the fitness difference is due to a difference in internal integration and 25 per cent due to the interaction with the environment' are not the best way forward. So from here on, please interpret 'internal selection' as meaning predominantly internal, and likewise for 'external'.

* * * *

Suddenly it begins to matter which type of creature we are dealing with. Although evolutionary theory, like all science, should be as general as possible, and although the idea of selection (internal and/or external) can be applied to any type of creature, there are some things in relation to which we have to be careful not to overgeneralize. And what follows is one of them.

We're mammals so let's start with a mammal embryo. It could be our old friend the microscopic horse; alternatively it could be the human embryo that, on a probabilistic basis, is growing inside a small proportion of my audience as they read these words. Equally, it could be a mouse or a deer embryo. All that matters is that it is a placental mammal.

What all placental mammal embryos have in common is protection from the elements. In this, they differ from many other types of animal, whose earliest developmental stages are 'out there', exposed to the vagaries of the physical environment, predators, and other threats. This is true, for example, of marine 'trochophore' larvae.

The superb protection from the elements afforded by the mammalian womb has a major effect on the balance of internal and external selective forces that act on the embryo. In this sort of situation, internal selection reigns supreme. As we know, many mammalian embryos die before they are born; human miscarriages, for example, are all too common, and doubtless the same is true of our various mammalian cousins. But this mortality is not caused by intrauterine predators. Nor can it be caused by harsh winters, except indirectly when the death of the embryo is caused by the death of the mother. What kills mammalian embryos, therefore, must be problems within the developing embryo itself, though problems with the umbilical connection to the mother are sometimes also a factor.

So we are back to the business of building bodies. In the case of the mammal embryo, and indeed of any other embryo in the true sense of the term (i.e. as distinct from a larva or a juvenile), building a body is its main job. Of course, it has to function well enough to stay alive in the process. Blood, for example, must flow to all the developing tissues. But there is no need to find a suitable microhabitat in a heterogeneous and mostly unsuitable environment. The mother has already provided it with the most secure microhabitat available. There is no need either to search for food. Again, the mother provides. So in this unique, highly protected situation, most of the selection that goes on is based on differences in organismic integration.

But what do we mean by 'organismic integration' in an organism whose very essence is a state of change? This is a very different scenario from the joint between the wing and the thorax of a fly that we considered earlier. That was integration of the *adult* organism. This adult, or indeed any other, has a very different – we could almost say 'opposite' – range of problems to deal with than an embryo. It does have to deal with the external environment and all the threats that that entails. But it does not have to transmute gradually into some very different kind of creature. Although the adult does indeed have to make another 'creature' – the sperm or egg that will, when they meet, form the next generation – this is a very different task from that faced by the embryo. An adult confines this process to its reproductive tract; the rest of the adult body is built for survival. And it doesn't change much over time. If a wing/thorax joint works well when the fly emerges from its pupal case and first takes to the air, it will probably still work well, injuries aside, later on in its life.

The embryo, in contrast, has a much lesser job to do on the survival front but a much greater job to do in terms of building a body. This is not something confined to a particular part of the embryo; rather, the whole embryo is, throughout its entire existence, transforming itself into something else. The embryo is the epitome of the phrase 'life flows'. Indeed, one way to picture an embryo is as a trajectory through multicharacter hyperspace. Even this is too simple because it is not just character values – like brain size – that vary as the embryo develops. Characters that were not there initially gradually come into being. We go from no brain to proto-brain to small brain to bigger brain.

Integration, in this sort of situation, comes in two forms. At any precise moment, the embryo must be sufficiently well integrated to function; to stay alive; to avoid the fate of miscarriage. But, when viewed over an extended period of time, the embryo must remain integrated in a different way. As well as the early embryo and the later embryo each having their own 'instantaneous' integration, there must also be a sort of 'temporal integration', so that the trajectory from one to the other works. And although I have set this up as a two-point comparison, this is just a starting point for thinking about the situation. In reality, every 'stage' must be compatible with every other.

This compatibility has a directionality to it. An earlier stage makes a later stage but not vice versa. So if there is any change in an earlier stage as a result of mutation in a gene that helps to control what happens at that stage, then not only must the instantaneous integration at the point when the mutation first takes effect be maintained, but also, if the embryo is to continue on its way later, the temporal integration between stages must not be lost. But there is a lot in this phrase 'on its way'. On its way to where? Well of course it is on its way to later embryonic stages, to juvenile stages and ultimately to the adult. But these may vary to some degree. After all, early embryos are not evolutionarily immutable. We know this because the early embryos of different mammal species are often similar but never identical. And the same applies to other groups.

All this simply means that embryos have to either stay the same or else change in a way where integration is maintained or enhanced. If integration is lost, the embryo dies. If integration is reduced, the embryo may live and continue to develop; but, other things being equal, the mutation causing the reduction will be removed from the population by natural selection. This last point is important, because we should never forget the probabilistic nature of evolution. This is a point that is deeply embedded in the subconscious of population geneticists. But developmental biologists tend to think in a different way: about changes, for example, that make the embryo 'inviable'. There is a risk, if we think in terms of viable and inviable embryos, of treating the effects of mutations as all or nothing, which certainly does not reflect reality. What we need to do is to unite a time-extended developmental view of organisms with a probabilistic view of the nature of evolutionary changes. We must take the strengths of each discipline and put them together. Only then

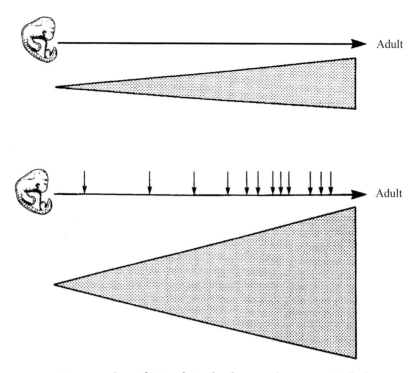

FIGURE 19 In evolution, later developmental stages typically become more different from each other than earlier ones for two reasons: the simple asymmetry of causality where early stages can affect later ones but not vice versa (top); *plus* the fact that it is easier to modify later stages (bottom). Vertical arrows indicate times of initial developmental expression of genes that alter in evolution; shaded triangles indicate increasing cumulative developmental divergence between two evolutionary lineages; larger triangle includes both 'asymmetry' and 'ease' effects.

will a form of evo-devo emerge that is truly mechanistic in both time frames.

*   *   *   *

Let's dissect this embryonic directionality, in which earlier stages make later ones, from an evolutionary point of view (Figure 19). One key question here is: how does the probability of a mutation in a developmental gene being advantageous or disadvantageous vary with the time of first expression of the gene concerned in the growing embryo? Do mutations in early-acting genes have greater, equal or lower probabilities of destroying that all-important temporal integration of the overall trajectory? Or, if you prefer to put it in a way that relates more to evolutionary change than to evolutionary conservation: do mutations in early-acting genes have a greater, equal or lower probability of causing improvements?

The conventional wisdom is that the earlier developmental reprogramming takes place, the more likely it is to be detrimental. If you change some fundamental early step, like the formation of the anteroposterior axis of the embryo, you are asking for trouble. Too many downstream processes are likely to be disrupted. The chances are that the embryo will die. But let's move forward in embryonic time to a much later stage when digits are being formed. Here, changes are not so likely to be lethal. Some humans have extra fingers and toes – 'polydactyly'. This is the result of a gene mutation that deflects the process of digit formation right back at its origin in the first couple of months of embryonic development – still 'late', though, in our comparison. But the new trajectory works reasonably well. If the developing digits were used to make something else further downstream, and their exact number was crucial, then the polydactyly mutation might be lethal. But this is not the case.

Although polydactyly is a genetic and developmental phenomenon, it is not an evolutionary one, at least not in our recent past. There are no human races whose members typically have six fingers on each hand or six toes on each foot. Instead, individuals are affected only very occasionally. But in the grand scheme of vertebrate evolution, changes in the number and structure of digits have been common. Some of the earliest tetrapods, it appears from the fossil record, had eight toes on each foot. And many modern mammals have a reduced number of digits – for example the horse. Also, the difference between a webbed foot and a 'free-toes' foot is probably due to mutations that affect the process of programmed cell death that normally occurs in the interdigital regions. So evolution can and does frequently modify later embryonic processes on which not too much depends in the way of generating other things further downstream. But in contrast, evolution almost certainly finds it harder to modify those crucial early stages on which all later events depend.

*　*　*　*

The story so far, if true, would mean that if we took two species of mammal – say a human and a horse – and compared their embryos at various points through developmental time, we would find that they looked most similar at the earliest stages and most different at the latest ones. And this is indeed what happens. This is what Karl Ernst von Baer observed in the early nineteenth century.[1] It was given an evolutionary interpretation by Darwin himself,[2] who asked how we could explain the fact of 'embryos of different species within the same class generally, but not universally, resembling each other'. He answered his own question by saying that it was explicable on the basis of 'descent with modification', and with many of those modifications only taking effect part-way through development, thus leaving the embryos more similar than the corresponding adults. 'Thus', he concludes, 'community of embryonic structure reveals community of descent.'

This von Baerian picture of diverging embryos, which Darwin accepted and gave evolutionary meaning to, has been questioned in recent years from several different points of view. In none of these cases do I think that our picture should be radically altered; that is, I believe that von Baer was broadly correct in his observations and Darwin broadly correct in his interpretations of them. However, the criticisms that have been made are instructive in certain ways, as we shall see. The central point in all the criticisms is that the standard picture of von Baerian divergence is too simple. Let's now look at the various aspects of this issue.

First, the very earliest embryonic stages may actually vary *more*, not less, than the immediately following ones. This leads to a bundle of developmental trajectories of different species within a class or phylum resembling not an ice-cream cone but rather a very asymmetric egg timer with its point of constriction close to its base. For many comparisons, this is a more accurate picture, and it is often related to the different kinds of egg environment that different embryos start in; for example eggs with very different amounts of yolk. A mammalian embryo supplied with nourishment from the placenta can, in its earliest stages, be quasi-spherical. In contrast, a bird embryo, which shares its eggshell home with a vast quantity of yolk, may have to start off as something more akin to a little disc curved across a small part of the yolk's periphery.

Second, von Baer and other nineteenth-century embryologists only looked at the embryos of a rather small number of species; and these constituted a non-random sample of the animal kingdom. The more species you look at, the more complex the overall picture becomes. How, for example, do you apply von Baerian divergence to the case of groups with indirect development, like flies? My feeling on this is that with regard to *embryonic* development – that is, the formation of little larvae within flies' eggs – von Baer's picture is still reasonably accurate. However, it can't be so easily applied to that later storm of developmental activity that breaks out during metamorphosis.

Third, it has been suggested by Leiden-based biologist Michael Richardson[3] that one of the nineteenth-century embryologists (Haeckel) 'doctored' his drawings, in the same way that Mendel is thought to have 'doctored' his ratios of pea plants. Well, even if this is true, which has been disputed, I don't believe that the doctoring was sufficiently extensive that the basically divergent picture should be discarded. It may be a little messier than some of the drawings suggested, but the cone (or egg timer) remains.

Fourth, we should be cautious, as ever, in inferring a mechanism from a pattern. Let me play devil's advocate for a moment. Suppose that my earlier assertion about advantageous mutation being 'easier' in later developmental stages than earlier ones is false. Suppose instead that 'evolvability' remains constant over developmental time. If this 'null model' were true, we could still end up with a pattern of von Baerian divergence among the ontogenetic trajectories of related species, simply because later stages will be affected by all changes whereas earlier ones will only be affected by a subset of them. However, although this simple and obvious asymmetry could, on its own, explain what we see, I don't believe for a moment that it does; rather I think that it is only part of the picture – see Figure 19. That is, I don't believe my own devil's advocacy. I suspect that the probability of beneficial reprogramming of development is strongly correlated with developmental time (with some complications that we'll get to later); but rigorous quantitative testing of this suspicion remains in the future.

Finally, biologists who work on marine groups characterized by a small planktonic larva – whether the molluscan trochophore or the echinoderm pluteus – tend to be less inclined to support von Baerian divergence as a general phenomenon than those who work on the embryos of terrestrial direct developers such as mammals or birds. This is quite understandable, of course. These small marine larvae should not be expected to behave in the same way as mammalian embryos. There are two good reasons for asserting this. First, because these forms are *larvae*, they need to make their own way in the world. They float around in the ocean rather than in the womb. Thus external selection will play a greater role here than in true embryos. Second, in at least some cases, only a small part of the larva is used to make the adult. So, just as the adult fly is made from its larva's imaginal discs, in many echinoderms the adult is made from just a small part of its larva – a part known as the 'rudiment'. So internal selection will play a less important role in the reprogramming of the development of most parts of the larva. This means that major changes in early development that occur in the divergence of different species, such as the evolutionary switch from a larva with feeding 'arms' to one without (because it has a supply of yolk) are to be expected. This particular form of developmental reprogramming has occurred in several separate evolutionary lines of sea urchins, as has been noted by the American biologists Rudolf Raff, Greg Wray and their colleagues.[4]

*   *   *   *

So it all comes back to the problem of deciding upon the appropriate level for generalization. In my view, if we get this right, a lot of apparent disagreements disappear. If we attempt to construct a 'law' that is applicable to all multicellular organisms, we will probably fail. Certainly neither von Baer's nor Haeckel's 'laws' are as general as that. Rather, we need to strike a balance between science's quest for generality and simplicity on the one hand and the messiness that actually prevails in nature on the other. I wish that I could remember who gave the wonderful advice 'seek simplicity, but distrust it'. This, to my mind, is exactly how biologists should proceed. We all try to build simple general theories and apply them to the widest range of taxa and phenomena imaginable. But then we look, one by one, at these taxa and phenomena and ask whether our generalization is justified. If not, we try one level down, and so on. Sometimes it turns out that our would-be universal law has not lost much generality because it is indeed OK one level down from where we started. In other cases, nature is less kind and we end up with a statement that is only true for a particular family (say) or for a particular short stretch of the whole of developmental time.

Let's see if this hierarchical view of the generality of theories can be made to work in the evo-devo realm. The following attempt to do this will take the form of a series of statements with, in each case, a comment on the level of generality at which I believe it is possible to apply the statement concerned.

'Adaptive evolution of development is caused by the three processes of mutation, developmental reprogramming and natural selection.' True of all multicellular creatures.

'The direction of evolutionary change is caused by the interplay between mutational/developmental bias and selection, and is also influenced in unpredictable ways by historical contingency.' True of all multicells, but probably with considerable variation from case to case in the relative importance of these direction-influencing factors.

'There are important, albeit stochastic, patterns in the 'ease' of achieving selectively advantageous reprogramming; these take the form of trends over the course of developmental time, but not necessarily simple ones.' Personally, I believe that this also applies to all multicells.

'There is one particular pattern that is found in cross-taxon comparisons, namely similar early stages giving way to progressively more different later stages as development proceeds.' It seems to me that this pattern is found, often with the egg-timer complication superimposed on it, in some big chunks of the living world but not others. The 'chunk' to which it does apply is not neat and tidy. It is perhaps truer of vertebrates than invertebrates, truer of direct developers than indirect, and truer of terrestrial organisms than aquatic ones.

'This pattern is caused by variation in the 'ease' of achieving changes, as discussed above, and by the simple asymmetry that early changes affect most stages whereas late changes only affect late stages.' This has the same level of generality as the statement immediately above; but we must add the proviso that the relative contributions of these two mechanisms to the observed pattern have yet to be determined.

'Evolutionary changes in the course of development lead in the direction of increasing complexity.' This is a good example of a statement that can be applied to some lineages but not to others, and that even in the former is only visible over very long periods of time. Many lineages do not increase or decrease in complexity much of the time – rather, they diversify within a broad level of complexity.

* * * *

These few examples help to illustrate what I mean by levels of generality. You can, no doubt, come up with additional assertions that you might wish to make about the evolution of development, and see if you can arrive at a defensible view of where they fit, in terms of which is the highest generality level at which they seem to be true.

It's probably worth making the effort to do this. Personally I don't want to list any more examples here. But I do want to sound a note of caution.

My friend and colleague Alec Panchen has said, in his book *Classification, Evolution and the Nature of Biology*,[5] that rather than looking for universal laws, as physicists do, biologists should look for what he calls 'taxonomic statements'. In other words, he is urging caution about extrapolating something that is true within one level (or rank) of taxon to a higher level where it may have exceptions. This is very sensible, and, in one way, I have no wish to argue with it. But I do need to make the point that in relation to generalizations about the evolution of development, descending the taxonomic hierarchy may not always be the right route towards restricted generality, when it becomes apparent that restriction is necessary. Other forms of restricted generality are possible. So, for example, if 'generalization X' applies to all species of direct developer and no species of indirect developer, then, given the irregular way in which direct development is scattered across taxa, we do not end up making a 'taxonomic statement' at all, at least in the normal sense in which that phrase is intended. After all, whatever this mythical 'X' is, it will apply to birds and mammals but not to most amphibians. Yet, going way out from our vertebrate starting point, it will apply to landsnails and slugs, but not to their aquatic relatives. So it is a 'statement of restricted generality', but not a 'taxonomic statement' in the sense of applying only to a particular clade.

12 Action and reaction

As I mentioned earlier, I started off my academic training in the field of ecology, moved sideways into ecological genetics for my Ph.D. and the following few years, and then moved sideways again into evo-devo. In Chapter 3, I discussed an accident that played a major role in initiating the latter sideways shift. In the best tradition of *Star Wars*, here, later, is the rather less accidental story behind the former shift – 'Episode 1', if you like.

By the time I reached the final year of my B.Sc., I had developed a particular interest in evolution. I wanted to conduct my final-year research project in that area – but how? My tutors were mostly ecologists, so they weren't adept at finding dinosaur bones. But that didn't matter, as I was (and still am) primarily interested in the operation of evolutionary mechanisms rather than the reconstruction of evolutionary history. (I have become more interested in history – both evolutionary and scientific – with age, as tends to happen to people, but it is still not my main focus of attention.)

So the task at hand was to choose an evolutionary project that was 'doable', and a species on which to do it. In the end, I came up with a project that involved comparison of shell shape between two populations of a species of pondsnail that inhabited very different types of environment. At first sight, this seemed like a wonderful 'adaptive scenario'; and this was in the days before that phrase took on a disreputable flavour and became associated with the derogatory expression 'story telling' – in the sense of plausible but not rigorously tested hypotheses.

The putative adaptive scenario went like this (Figure 20). One population, of a few hundred snails, lived in a pond that was relatively

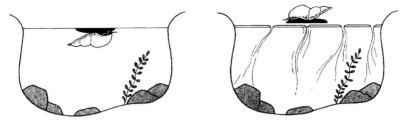

FIGURE 20 A tale of two ponds: pondsnails living as aquatic (left) or quasi-terrestrial (right) creatures.

free of vegetation. There was the usual lush pondside flora, but the water surface was essentially plant free. The water was clear, and if I looked hard I could see into it and could make out some snails crawling over submerged stones. Others had the peculiar habit of hanging upside down from the surface film. No snails could be seen out of the water, for example climbing up any of the pondside plants. So this population of pondsnails did indeed live, as you would expect, in the pond.

The other population, again of a few hundred individuals, inhabited a very different pond. Although of roughly similar size to the other, this pond was completely overgrown. Only about five per cent of the water surface, and indeed perhaps less than that, was visible. The rest was matted with the flat leaves of a water plant that grew profusely and provided a platform sufficiently stable that creatures as large as frogs, an order of magnitude heavier than my snails, could sit on it without risk of falling through. Here, most of the pondsnails, most of the time, could be seen gliding slowly over the surface of the leaves. I suppose they were feeding on the coating of epiphytic algae that such leaves often have. But whatever they were doing, they were living in an essentially terrestrial environment, albeit a rather unusual one suspended over water.

We have already encountered the problem of desiccation that is faced by most terrestrial invertebrates. But it is not normally a problem for their freshwater counterparts. How can you risk dying of water loss if you are surrounded by water? Not very easily. But real risks in nature are not determined by textbook names. You may be called a 'pondsnail', but if you live as a landsnail you will face all the same problems as landsnails, of which desiccation is one of the most prominent. Perhaps the problem will not be quite so severe for our out-of-water pondsnails because the humidity level immediately above a pond will be higher than that above, say, an area of grassland. But a high humidity is no substitute for liquid water; it will reduce the rate of desiccation a bit, but it will not entirely remove the problem.

Some snails have a thing called an operculum, which is a sort of door that can be pulled across the hole in the shell (the 'aperture') after the snail has withdrawn inside. This reduces water loss; it makes life harder for would-be predators too. But my species of pondsnail is not blessed with such an adaptation. It has no 'door', so the aperture remains permanently open, and provides the main site of water loss. Quite a problem for a pondsnail that is pretending to be a landsnail.

But actually, this is where things begin to get really interesting, because there was a difference in shell shape between the two populations that was so large that it was apparent to the naked eye without the need for measurements. (I did, of course, make the relevant measurements later to quantify the difference.) The 'terrestrial' population consisted of individuals that had unusually small apertures. In contrast, the individuals in the 'normal' population had quite large apertures. Yet the overall lengths of the shells were the same. So this was a difference in shape rather than size. It was probably caused by different relative rates of elongation and widening of the 'tube'. (A typical snail shell is just a coiled tube, though you need to mentally unwind it to see the tube rather than the coil.)

So, this looked like a result of natural selection. The scenario that sprang to mind was: my two populations had arisen from a common ancestral population sometime in the past; they had migrated to different habitats; those habitats had very different environmental conditions; one set of conditions was rather odd and caused the snails to have a more terrestrial existence than is typical for their species; natural selection thus came into play because individuals that lost water more rapidly – those with bigger apertures – suffered greater mortality than others; so the average aperture size of the population gradually decreased.

I became quite excited by this. Maybe I had found an example of evolution in action to rival the peppered moth story; maybe this would get into the textbooks. But my excitement was short-lived. Like many adaptive scenarios, this one turned out to be complete nonsense. It didn't take long to discover this.

What is missing from the story so far is any information on the genetic basis of shell shape. I didn't expect this to be as simple as, for example, the single gene that determines that a moth should have melanic, as opposed to pale, 'peppered' colouration. It has been known for a long time that the genetic basis of continuous variation (as in shell shape) is usually more complex than that of its discrete equivalent. And the possibility of partial heritability, wherein not all the variation is genetically based, rears its ugly head.

So I did a very simple, and very imperfect, experiment. All science is done 'against the clock', as we all have finite working lives. In the context of a final-year undergraduate project this clock effect assumes even greater importance than usual, because the project must be devised, carried out and written up all within the confines of an academic year – about nine months. In a species of snail whose generation time is twelve months, this imposes major restrictions on experimental design.

What I did was this. I filled two tanks with water from the overgrown pond whose population had the smaller apertures. Then I put samples of about twenty or thirty baby snails into both tanks. In one case, the snails came from this same pond, so they were growing up, in the laboratory, in a small sample of their original habitat. In the other, the snails came from the 'normal' habitat, so they had effectively been 'transplanted'.

Now looking back at it, this was a truly dreadful experiment, and it was sheer good luck that I got an interesting result out of it. The tanks were too small; the samples were too small; the timescale was too short; the water plant was missing; and so on. But here is what happened. After some months of growth, I measured the aperture sizes relative to the overall shell sizes in both tanks. The distributions of 'relative aperture size' in the two samples were identical. The transplanted population had shifted away from its parent population until its average shell shape coincided with that of the population in whose water it had been reared. The difference in shell shape between the populations was not genetically based; it was caused by a direct effect of the environment on development. This was not evolution at all. Rather, it was an example of what is known as phenotypic plasticity – deflection of the developmental trajectory by the environment. Exactly what aspect of the environment I never pinned down, though one subsequent experiment suggested that the concentration of calcium in the water might be a causal factor.

Many other outcomes would have been hard to interpret. If the transplanted population had not altered, it might have indicated a genetic basis of the variation. But equally, since the transplanted individuals were baby snails rather than eggs, such a result would also have been compatible with a hypothesis that the variation was determined environmentally, but at some crucial early stage in development before the transplant took place. And a partial shift in aperture sizes would have been even harder to interpret – maybe a combined genetic and environmental effect, but not necessarily. In this case, luck didn't favour the 'prepared mind', it just happened.

Although I was pleased to have got a clear result, I was disappointed that the adaptive scenario that I was initially so excited about turned out to be false. Before my move to evo-devo many years later, I took the view, common in evolutionary genetics circles at the time, that plasticity was just an annoyance. It meant that the observed variation was not inherited and so was irrelevant to evolution. But this, as I subsequently learned, was a seriously misguided view. Plasticity *is* important in evolution, despite its lack of inheritance, as we will shortly see.

* * * *

And now, from snails to flies. There is a famous mutation in fruitflies called 'bithorax', where the little balancing organs that stick out like miniature drumsticks immediately behind the wings are transformed into a second pair of wings. In fact, the whole segment concerned has been altered – from a third thoracic segment to a duplicate of the second. This is a rather drastic form of developmental reprogramming, which belongs to the category that is called 'homeotic' – the right thing in the wrong place.

Fruitflies are, of course, much better known genetically than snails, having been a favourite workhorse of geneticists for about a century. So we know a lot about the mutation that produces this bithorax phenotype. We know which gene is involved; and we know where to find it – about half-way along chromosome 3. But the same phenotype can be produced in different ways. Sometimes a phenotype that can be produced by a gene mutation can also be produced by the environment. And since adult phenotypes are the end results of development, what this means is that an environmental factor can 'imitate' the developmental effects of a mutation. This is referred to as phenocopying.

The inspirational Edinburgh-based geneticist C. H. Waddington, whom I mentioned earlier, did some interesting work on a phenocopy of the bithorax mutation.[1] This can be produced in some genetically 'normal' flies by exposing the eggs to ether vapour. This effect, like the reduction in the aperture size of my snail shells, is an example of phenotypic plasticity. I suppose the best way to look at it is that phenocopying generally is a subset of plasticity. If a certain alteration to development can be achieved either by mutation or by an environmental factor, then the environmentally induced variant can be called a phenocopy. If, in contrast, the environment can alter development in a way that no gene mutation is known to do, then it is not a phenocopy, because there is nothing to copy; but it is still plasticity.

Actually, the term phenocopy is probably best avoided. It is a hang-over from 'genetic imperialism': the real thing is caused by a gene; the same thing produced by an environmental factor is a mere copy. I don't think that this asymmetry of terminology is helpful.

What Waddington discovered, in a series of experiments done in the 1950s, was that if he (a) exposed eggs to ether vapour, (b) found that this transformed some flies to bithorax, others not, and then (c) bred the next generation only from transformed parents, the proportion of flies that had their development transformed on exposure to the same amount of ether vapour increased. When he continued the experiment for many generations, it increased further. And then some flies even appeared in the experimental population that developed the bithorax phenotype without having been exposed to ether at the egg stage. That is, a feature that initially needed a particular environmental stimulus to induce it eventually appeared without that stimulus – a phenomenon (genetic assimilation) that we examined briefly in Chapter 2.

*　*　*　*

Genetic assimilation is a special case of a more general phenomenon: selection on 'reaction norms'. Let's have a look at what this curious phrase means, and then move on from fruitflies to butterflies for an example.

If you take any particular kind of creature, and rear it in a series of different environments, you are likely to get different results. Sometimes these will be broadly predictable, other times not. An example of the former is variation in body size across a range of environments where food varies in abundance – less food, smaller size. Waddington's experiments provide an example of the latter. Who could have predicted that environments imbued with ether vapour would produce this particular drastic alteration to the course of development? These examples also show that such environmental effects can be continuous or discrete.

The pattern of variation of a phenotypic character in response to variation in some environmental factor is referred to as its 'norm of reaction', its 'reaction norm' or its 'developmental reaction norm', the last of these often being abbreviated to DRN. I find these terms a little odd, but I suppose that the best way of interpreting them is that the environment is 'acting' on the developmental system, and the system itself is 'reacting'. Quite what the 'norm' is supposed to convey I'm not sure. Perhaps it is a sort of variable phenotype equivalent of the old fixed-phenotype concept of a 'wild type'. Anyhow, the terms are too well established for it to be sensible for me to try to replace them with new ones. But since I don't like them much, I'll depart from my usual preference for full words and use the abbreviated form, DRN.

The American biologists Carl Schlichting and Massimo Pigliucci have done much to develop the DRN concept and to emphasize its evolutionary importance – for example in their book[2] *Phenotypic Evolution: A Reaction Norm Perspective.* On the cover of this book is a picture of two butterflies with different sizes of eyespots on their wings. This picture comes from a study undertaken by the Leiden-based expatriate Englishman Paul Brakefield[3] on a species of butterfly that we'll just call the 'African brown'. As you'll gather, I'm generally trying to avoid Latin species names. Anyhow, it belongs to the 'browns' family and it lives in Africa, so this label will do fine. Paul Brakefield and his colleagues have been working on this species, and in particular on variations in its eyespots, for a considerable time. They have amassed a wealth of information, a small sample of which follows.

The eyespots, which some biologists think serve an anti-predator role, are partly determined by the butterfly's genes and partly by the environment. So they can be altered (in size, number and intensity of pigmentation) both directly, within a generation, by the environment altering their development, and indirectly, over many generations, by natural or artificial selection that increases the frequency of genes that give the 'desired' kind of eyespots.

In nature, one of the most conspicuous elements of the overall pattern of variation is the difference between wet season and dry season forms, with the former having much larger eyespots than the latter. This difference falls under the heading of polyphenism. The wet and dry season forms are produced by the direct effect of the climate on the developmental mechanisms that underlie eyespot formation. But the Leiden group showed that it was also possible to select for larger eyespots in laboratory experiments extending over several generations.

This is not the same as selecting for a fixed character. Both the pre- and post-selection butterflies developed different sized eyespots depending on the temperature at which they were raised – the higher the temperature, the larger the eyespots. So instead of shifting the mean value of a character that is genetically fixed in each individual, these artificial selection experiments on eyespots were shifting the appropriate DRN. At any particular rearing temperature, the butterflies selected for larger eyespots had bigger spots than their unselected counterparts. This was a genetically based difference, while the naturally occurring seasonal polyphenism is not.

* * * *

This work on the 'African brown' involved elegant and well-designed experiments that produced informative results. It has become, in its way, a classic case study. Yet there is a danger that, despite its beauty and clarity, it might be simply regarded as just another in the increasingly long line of famous case studies, following on from Darwin's finches, melanic moths *et al*. But thinking of it in this way would be a mistake. It is more than just another case study. It, and other studies like it on other creatures, have an impact that goes far further than, for example, the studies on industrial melanism.

The reason is this. Some DRNs are dramatic, as in the case of Waddington's flies. Others are less so, but probably more relevant to evolution in the wild, such as the DRN in butterfly eyespots. But the important point is that this small number of examples should not be taken to mean that DRNs are rare and that most characters are, in contrast, entirely fixed by the genes. Although we cannot yet attempt a quantification of the relative proportions of developmental variation in natural populations of all creatures that (a) have an environmental component and (b) are totally determined by the genes, my guess is that the former outnumber the latter. If so, then the *general* way that natural selection acts is by modifying DRNs; and those few 'fixed' characters simply constitute the special case (or subset of cases) where the DRN is flat. If that is a reasonable view, then the standard neo-Darwinian concept of selection acting on fixed genetic variants is subsumed in a wider view. You might almost want to call this a paradigm shift.

There is, however, a caveat. We are making an implicit assumption here, and, as ever, it is better to make assumptions explicit, so let's get it out of its hiding place and have a look at it. We are assuming that variation in DRNs is at least partially heritable. This need not necessarily be the case. It may be that some DRN variation is highly heritable, while some is not.

Notice that this is all beginning to get quite conceptually complex. In the case of a partially heritable DRN that is acted upon by natural selection, the situation can be summarized as follows:

> The value of character X is not fixed among the members of a population. Rather, it varies. Within a single environment, the variation is partially heritable – some of the differences between individuals are due to different genes, some to different microhabitats, amounts of food, etc. In some comparisons between two randomly chosen individuals, their difference may be mostly for genetic reasons, while in other such comparisons it may be mostly for environmental reasons. Over the whole population, the proportionate genetic contribution to the variation can be measured by the 'heritability'.

But DRNs then enter the picture either when the population migrates to a different environment or when the conditions in its initial environment alter significantly, or, perhaps most interestingly, when the population spreads to occupy a wider area which is heterogeneous in that parts of it have quite different conditions (of temperature, food supply and so on) from other parts. In this case, if the

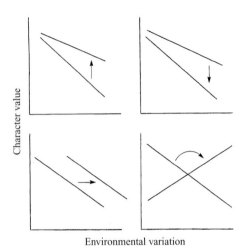

FIGURE 21 Types of evolutionary change (arrows) in developmental reaction norms (lines). Clockwise from top left: flattening, steepening, inverting and shifting.

Environmental variation

DRN is suboptimal in the sense that the interaction between the genetic/epigenetic program for development and the environmental conditions that prevail produce a phenotype that is not maximally fit, then selection will increase the frequency of genes that produce a system that interacts 'better' with the prevailing conditions. Exactly what selection does to a DRN in such a case will vary. Possible results of this process include: flattened, steepened, shifted and inverted DRNs (see Figure 21). And if DRNs are non-linear, which is likely to be the rule rather than the exception, then more complicated changes will be possible.

Now I don't blame you if you are beginning to rebel against this picture. I'm almost beginning to rebel myself. This is partly because I like simplicity, and the situation described above is far from being simple. But it is also because some of the distinctions I have made are a bit questionable. In particular, in describing within-population, within-environment variation in non-DRN terms while describing differences in mean values between subpopulations inhabiting adjacent but different environments in DRN terms is a bit unsatisfactory, especially given that one intergrades into the other. When does a microhabitat become a habitat? It depends on lots of things, including the size and mobility of the creatures.

But in some ways, the deficiencies inherent in this kind of picture don't matter. The fundamental point that I am trying to get across in this chapter, denuded of all terminological difficulties and the peculiarities of particular types of creature, is a very simple one. It is that phenotypic plasticity should not just be thrown out of the window by evolutionary theorists because it is not inherited. If you take one step back from the non-inheritance of 'plastic' variants, you will probably find that the pattern of plasticity is itself inherited, or at least partly so, and that there is variation for it in a 'typical' population.

*   *   *   *

It is important to relate all this back to the concept of developmental reprogramming that I introduced in Chapter 7. I said then that there was a constant interplay between the genes and their direct and indirect products, that is, an interplay between what can be called the genetic and epigenetic programmes; but also that this interaction could itself be altered by the environment. In the present chapter, we have looked at a couple of examples of such environmental influences, and that has led us to the developmental reaction norm. I think it is appropriate at this stage to re-emphasize a point made earlier, namely that environmental influences should not be considered as part of 'reprogramming'. It is reasonable to think of the organism as carrying within itself a programme (or programmes) for its own development. But it is hardly reasonable to think of a habitat as having a 'programme' for the development of organisms that live there. So it is still the case that the ultimate cause of developmental reprogramming is mutation, but in some cases mutations result in the reprogramming of development in one environment but not in another, or result in different kinds or amounts of reprogramming in different environments. These are the mutations that alter the developmental reaction norm.

13   Evolvability: organisms in bits

The history of life is a mixture of stasis and change. We can recognize both of these in any particular interorganism comparison. Take, for example, two different species of vertebrates, such as ourselves and the horse. Stasis is apparent everywhere. We both have vertebrae, ribs, limb bones and so on. Furthermore, we both have limb bones that follow the same basic pattern – a single long bone in the upper limb but a pair of roughly parallel long bones in the lower limb. One way of describing these similarities, which we have both inherited from a common ancestor, is as a common body plan. Another way of describing them, which can be used more widely, is homology. We say, for example, that the femur bone in the horse's hind limb is homologous to the femur bone in our own leg. We'll get to grips with homology to a greater extent in the next chapter, but all we're really saying, in this example, is that the last common ancestor of humans and horses already possessed a femur, and that we both inherited modified versions of it.

But change is everywhere too. Not a single one of all our 200-plus bones is identical to its horse counterpart. Different sizes and shapes are not just the norm, they are universal. Compare what you like: skulls, limbs, ribcage and so on. The conclusion is the same. Evolution has left nothing untouched.

If we could travel back in time and locate, precisely, the last common ancestor of human and horse, and compare it with either of its present-day descendant lineages that we are considering, the situation would be the same: same overall body plan, same set of homologies, but different shapes and sizes (and sometimes numbers) of the relevant components. Maybe in some characters the lineage leading to the horse has altered more from the common ancestral state, while in other characters the lineage leading to humans has altered more.

The same is true if we conduct a similar thought experiment with two randomly chosen members belonging to another phylum or class. Take grasshoppers and moths, for example, in the insect world. Both have the characteristic arthropod exoskeleton that is fundamental to this particular body plan. Both have three pairs of legs. Both have compound eyes. The hindlegs of a grasshopper are homologous to the hindlimbs of the moth, despite the gross enlargement of this pair of legs that characterizes the grasshoppers and their allies (such as crickets).

What we are seeing here is that some things are more 'evolvable' than others.[1] That is, some things are in some sense 'easier' to change. We looked at one side of this coin in Chapter 11 when considering the difficulty of altering early embryonic processes because of the dangers of knock-on problems later in the sequence of developmental events. But let's now look at the other side of the coin – features that enhance rather than restrict 'evolvability'. It follows from the argument about the problems inherent in early-effect changes that later-effect changes are generally more evolvable. But let's try to get a little more specific. Let's ask the question: what particular features of organismic design increase the probability that evolution will find a way to produce advantageous modifications?

*   *   *   *

One such feature that has been much discussed of late[2] is 'modularity'. Although the process of development is normally continuous in both time and space, there is some 'clumping' of activity in both of these dimensions. So we can recognize, and give names to, particular temporal 'stages' such as blastula and gastrula; and also to particular spatial 'modules' such as limb or eye primordia. The basic idea behind 'modules' is that they represent quasi-autonomous parts of the developmental system that perhaps can be changed without disrupting other things. Whether this is true, however, depends not just on how autonomous these bits of the embryo are, but also on the extent to which the genes used in their development are used elsewhere.

One of the most important forms of 'modularity enhancing evolvability' is 'duplication and divergence'. This occurs both at the morphological level of large body parts, like limbs or vertebrae, and at the molecular level of the gene.[3]

'Duplication and divergence' is a well-trodden evolutionary path, and there is a good reason for this. Evolution is continually faced with the problem of how to keep an existing operation going without interruption, while at the same time designing something new. If evolution could take 'time out' from playing the survival game, it would be a very different process from the one it actually is. But it can't. Every single generation in an evolutionary lineage extending over millions of years must provide an efficient survival vehicle; if not, the lineage stops in its tracks and is replaced by the descendants of others. Experimenting with new designs is highly dangerous in this kind of 'treadmill' situation. However, there is an exception to this general rule.

If an organism has multiple copies of any particular module, then it's just possible that one or more copies can be 'experimented with' while the other copies perform the original function well enough on their own for the organism to survive. There are numerous persuasive examples of this process, and it is probably widespread and responsible for the appearance of many evolutionary novelties. Let me give just two examples, one each from the morphological and molecular realms.

First, arthropod legs. The minimum number of pairs of legs needed to walk or run in a reasonably stable way is one. Well, perhaps we're a little biased here, being human. Nevertheless, thousands of species of birds are bipedal like us, so one pair really will suffice, and not just for humans. However, even if we err on the side of caution and assert that a smaller, differently designed creature like an arthropod needs a minimum of two pairs of legs for stable land locomotion, we can see that all the vast array of arthropods have more than this: three pairs of legs in insects, four in arachnids, 'many' in crustaceans and 'very many indeed' (sometimes more than 100) in the appropriately named myriapods. Given that there is an excess of legs over what is required for locomotion, surely some could be modified into other things? Well, yes, and there are lots of cases where this has clearly happened. Many arthropod mouthparts are smallish paired appendages that have been derived from 'spare' pairs of legs in an ancestor. The centipede's impressive poison claws, which we encountered in Chapter 5, have probably also been derived in this way.

However, while this process of divergence from replicated, initially identical structures is an evolutionary reality, we should not feel too complacent about our understanding of it, because although we can explain some features of organisms in this way, we have not yet developed any predictive ability. Centipedes, for example, would appear to have a particular surplus of legs, and yet the 'longest' ones, with 191 pairs, have not specialized any more of them into other things than the shortest, which have only fifteen pairs.

Now to the 'Hox' genes, a group of developmental genes that control the patterning of the anteroposterior axis in animals. It appears that the very first animals had just a single Hox gene. But as animal evolution proceeded, this gene duplicated on many occasions so that most present-day animals have multiple Hox genes. This has allowed divergence in the stretch of the main body axis over which each Hox gene exerts an influence. Some specialize in patterning the front, some the middle, some the rear. As in the case of limbs, the evidence for functional divergence of replicated 'modules' is persuasive. But again, we have little predictive ability. Centipedes should need fewer Hox genes than insects or crustaceans because the amount of anteroposterior patterning is proportionately much less – many of the segments are virtually the same. But they don't have fewer. Also, in vertebrates, which have the largest number of Hox genes, fish hardly need almost double the number that we humans have – but they have them anyway.

I suppose that we shouldn't expect too much. After all, evolution is an odd mix of stochastic and deterministic processes, with a fair number of one-off accidents thrown in along the way. Such a process is never going to be predictable in any reliable manner. So perhaps we should be rejoicing at how much we can explain, not bemoaning our predictive limitations.

* * * *

Modularity and the divergence of replicated modules are hot topics. They are old morphological themes that have struck a chord with new molecular discoveries. They seem to apply at both levels. Few biologists seriously object to these themes as ways of generating evolutionary novelty. But now I want to make a major shift to consider another possible influence on evolvability. This is not so much a shift from the sublime to the ridiculous as from the accepted to the speculative; or from the topical to the neglected (in the sense of being buried in those old zoology texts that you have to blow the dust off before opening).

This ancient topic that I want to exhume is a spectrum of possibilities that generally speaking has no name and has to be identified by the labels that are applied to its two ends: 'specialized' and 'generalized'. If you are able to find one of those dusty old zoology texts and look up these terms in the index you are almost certain to find them. In many modern equivalents, you will not. Is their demise premature? Should they indeed be consigned to the dustbin of undefinable and unquantifiable speculation? Or might they too have some potential for reincarnation in this new evo-devo era?

To be honest, I'm not sure of the right answers to these questions, but I think that a brief airing might not come amiss. At worst, it could waste a couple of pages and attract some flak; at best, it could stimulate some new lines of enquiry that might just go somewhere useful.

So, how do we define the undefinable? What do we mean if we describe an animal as 'generalized' or 'specialized'? The best way forward is probably through examples. Which is the more generalized mammal – a shrew or a giraffe? Most people would choose the former. Which is the more generalized mollusc – a snail or an octopus? Again, the former would be the more popular choice. Which is the more

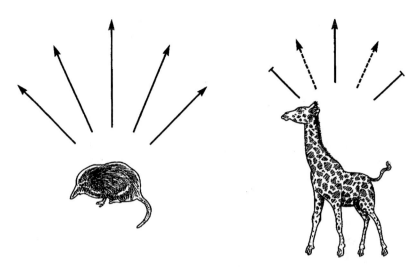

FIGURE 22 Possible inverse relationship between degree of specialization and future evolutionary potential. Solid arrow, easy; dashed arrow, difficult; blunt-ending lines, impossible.

generalized bird – a gull or a penguin? Yet again, the former would be the victor in any vote.

What is it, then, that is at the back of our minds when we make these decisions? I think it is something like this. Some developmental systems have not yet elaborated a basic body plan much in any direction. The shrew, for example, might be described as a fairly 'standard' mammal. The giraffe, in contrast, has elaborated the mammalian body plan in a definite direction – upwards. The same sort of contrast can be made in my mollusc and bird examples, except that the directions taken by the more specialized of each pair – towards intelligent manoeuvrability or streamlined flightlessness – are very different from the direction taken by the giraffe.

It may be that it is harder to find ways to reprogram development without loss of fitness in those more elaborated designs. Perhaps the very fact of choosing a particular kind of elaboration reduces the likelihood of being able to switch directions later. Perhaps realized evolution and future evolutionary potential are negatively correlated (Figure 22). Personally, I think there is probably something of merit lurking in these ideas, but I do find all the main concepts hard to pin down. Perhaps there will be a case for exhumation of the generalized-to-specialized continuum, but not quite yet. The way forward, if there is one, is not yet sufficiently clear.

*   *   *   *

Finally, I'd like to make one brief comment on another player in the evolvability game. This is the 'unmasking' of variation. Selection, the blind watchmaker,[4] cannot see through masks. Waddington's use of ether vapour to expose otherwise hidden genetic variation was an early example of a particular type of unmasking. But, after something of a gap, this idea of unmasking variation is making a comeback in a new and different way. It seems that some genes suppress variation in lots of different characters. If these genes mutate to a state in which their suppression no longer works, there is an outbreak of variation.[5] Work has really only just begun in this area and it is too early to say how important these 'unmasking' mutations are. But they are certainly worth further study.

After several chapters about ideas and mechanisms, let's return to history, and in particular to phylogenetic history, a knowledge of which provides a firm foundation for studies in evo-devo. If we know what the pattern of relatedness among different types of animal is, then we know what has evolved into what. So we can deal with real rather than hypothetical evolutionary changes in development. We can, for example, avoid attempting to explain how segmented worms turned into segmented arthropods, which many earlier evolutionists spent much time on, because recent advances in our understanding of phylogeny make it clear that no such evolutionary transition ever took place.

14 Back to the trees

Why so little phylogeny – evolutionary trees, that is – until now? Why, if the pattern of relationships among organisms provides a firm foundation against which to investigate the evolution of development, have I left it so late? When I wrote my last book[1] – *The Origin of Animal Body Plans* – I put the 'trees' chapter near the front; yet now it is near the back. This is not simple inconsistency. Rather, it's intentional, and now it's time to explain the reason behind this design.

Although I started off by contrasting three types of trees – adult, embryo and life-cycle trees – only one picture of evolutionary trees has appeared between then and now (in Chapter 5). How have we been able to proceed so far with so few trees? It all comes back to the difference between natural history and natural philosophy, and to the nature of the whole evo-devo endeavour.

The disciplines of population genetics and palaeontology contrast markedly in relation to their 'ahistorical' and historical natures. Population genetics investigates general mechanisms that should be applicable across the whole of evolutionary time, while palaeontology focuses on particular historical sequences that, if you move backwards or forwards in time, are replaced by entirely different ones. This is a bit of an oversimplification, because palaeontology has spawned general theories of its own, such as species selection, and population genetics includes some case studies that have a definite historical dimension. Nevertheless, the general contrast is a reasonable one.

How can evo-devo be classified in this respect? Is it natural philosophy or natural history? Does it seek general theories or historical reconstructions? In my view, it seeks both. It has a much more 'hybrid' nature than those other two evolutionary disciplines. Some of its proponents veer one way, some the other. But there's no conflict here; indeed I regard this hybrid nature as a strength rather than a weakness.

As you'll be aware by now, I am one of those proponents of evo-devo who incline more to the 'general theories' side. And that is why we have been able to come so far with so few trees. The central points of the last few chapters have all been about general mechanisms. Their applicability is intended to be very broad: essentially to the whole of the multicellular world. Developmental reprogramming, developmental bias, internal selection, reaction norms and modularity are concepts that transcend the confines of individual clades. So trees weren't necessary. But now they are.

The reason for this is that in the next chapter we will make another mental leap – into the world of molecules; and specifically, the molecules that make development happen in a variety of different creatures. This is a hugely exciting area characterized by an explosion of recent research. But some of the results that have emerged from this research have turned out to be puzzling and quite difficult to interpret. In the end, it is impossible to make much headway in understanding them without knowing something about patterns of relatedness. Hence the placing of our discussion of trees at this at-first-sight illogical point.

There are three things that we need to know about trees: first, how to construct them, and how the methods of doing this have changed over the last half-century; second, how to interpret them, and how to appreciate their limitations; third, how to apply them in the attempt to understand the strange recent results of what you might call the 'comparative developmental genetics' wing of evo-devo. (Maybe heart would be a better word than wing here.) I'll try to take all three of these on board in the course of this chapter and the next.

* * * *

So, how do we make a tree? This is another of those deceptively simple questions. Let's concentrate, at least initially, on the simplest possible situation, namely the relationships between three species. If you only have a single species, there is no relationship to explore. If you have two species, all that needs to be said is that they are related. All living organisms are related, so whichever two you choose, their relatedness can be represented by a simple bifurcation, in other words a two-pronged fork. You might want to add some temporal detail, of course, like when the parting of the ways took place. But in terms of who is most closely related to whom, there is no question to be addressed. This sort of question first arises when we have three species. Unless evolution produced all three as simultaneously appearing daughter species of a single parent (not impossible), then two of them are more closely related to each other than either is to the third.

There are two ways to approach this question: using real creatures or abstract symbols. I'm going to take the former approach. This is partly because I once attempted to read a book on tree construction where the whole explanation was based on the use of A, B, C etc., and their possible relationships; and while I didn't actually fall asleep, it was a close thing.

I'll range through quite a few creatures in the course of this chapter, but I'll start with a shark, a horse and a dolphin. If you're one of those folk who prefer the abstract approach, just think of them as S, H and D. Now we have to try to do something that is rather difficult. We have to suppose that we are our own cave-painting ancestors of many thousands of years ago. In other words, we have to imagine that we know very little biology. We have never dissected anything, except perhaps for culinary purposes, so we have little knowledge of internal anatomy. We judge creatures essentially from their external appearance. We have seen lots of horses, and we have seen the occasional dead shark and dolphin washed up on the beach.

'Cave-painting-man' was sufficiently recent, in the grand scheme of things, that the sharks, horses and dolphins of the day would have been very similar to their present-day equivalents. Indeed, we can, for our purposes here, assume that they were identical. Our task, in the first biology lesson ever, is to arrange them in a tree that indicates their pattern of relatedness. We would perhaps react to this task in the same way as a teenager of today would react to the task of

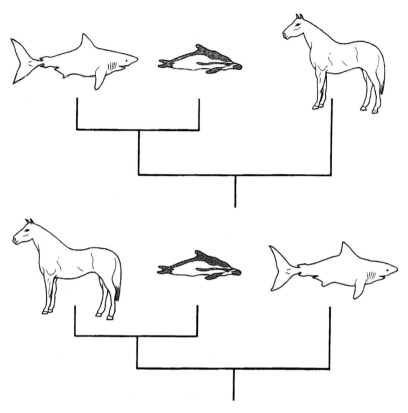

FIGURE 23 Two possible patterns of evolutionary relationship between sharks, horses and dolphins. Top: 'Caveman's tree'. Bottom: Present-day tree.

adding the numbers 2 and 5. That is, we would see the task as insulting our intelligence because the answer is obvious. On the basis of external appearance, it seems clear that sharks and dolphins are the closest relations out of the three, so we would draw the phylogeny as shown at the top of Figure 23.

Back in the present, with pretences of being our own ancestors cast aside, we would draw the lower picture in the figure. Why? That is, what knowledge would cause us to disregard the obvious information about overall body shape that is staring us in the face? This question can be answered at two different levels. If we don't want to go into things too deeply, we can simply say that it is well known that horses and dolphins are mammals, while sharks are fish. That in itself is sufficient justification for drawing the lower, rather than the upper, picture. But we might want to delve a little deeper and ask how it is that we know that dolphins are mammals. In fact, there are many clues to mammal ancestry in the dolphin's body. Mammals get their name from their mammary glands, and dolphins, like other mammals, possess these. They also have large mammalian brains, which is why they are so much more intelligent than any fish. And if you dissect a dead dolphin's fin, you will find the skeleton of a hand rather than the series of cartilaginous rods inside the fin of a shark.

I've picked this example to start with, because there is no doubt about the truth. We know that the caveman's tree was wrong and our present-day tree is right. The parting of the ways between fish and mammals happened many millions of years before the divergence of the lineages leading to the horse and the dolphin. Nevertheless, the caveman's tree was not based on a complete lack of bodily features to assess in order to make a guess at the pattern of relationship. In fact, this tree was the only sensible assessment of the evidence that was available at the time – that is, overall external body form.

There is a general problem lurking here that becomes much more severe when we don't know in advance what the answer should be. This is the problem of different characters suggesting different trees. We know enough, these days, about the vertebrates, to give precedence to the presence or absence of mammalian characters over the general shape of the body in deciding which tree is correct. But in other cases of character conflict it is much less clear which characters to 'believe', as pointing us towards the true pattern of relationship, and which to regard as misleading. This problem can occur at any level, and it is worth examining it at two very different levels – those of the family and the phylum. We'll take them in that order.

The family that I want to focus on is our own – the Hominidae. In particular, I want to examine the pattern of relationship of just three animals within this family – gorillas, humans and chimpanzees. If you ask the proverbial person on the street to draw an evolutionary tree for these three species, you will probably be handed a piece of paper with a picture like that shown at the top of Figure 24. Again, as in the caveman example, rational thought underlay the production of what turns out to be the wrong answer. The typical person on the street knows little of internal anatomy and even less of molecular biology. So such people are forced to base their decisions on what they know of the external appearance (and perhaps also the behaviour) of the animals concerned. Gorillas and chimps are both hairy forest dwellers of (comparatively) low intelligence. We, on the other hand, are almost hairless, inhabit cities, and have invented refrigerators and spacecraft.

The correct tree, which has only emerged in the last few years, is shown below the incorrect one in Figure 24. This is another case of character conflict. Hairiness and intelligence point to one tree, but other characters, which biologists have ended up agreeing should be given preference, point to the other. Some of these latter characters are molecular ones. If you determine the sequence of building blocks in either a gene or the corresponding protein (nitrogenous bases and amino acids respectively) in ourselves and in the gorilla and chimp, what you will typically find is that our sequence and the chimp's are the closest match. The differences are very small, though, for any of these comparisons, and so they should be interpreted with care. But by now sufficiently many comparisons of different genes and proteins have been made that the true pattern of relationship – i.e. the bottom picture in Figure 24 – has become clear. When we add in the fossil evidence too, it looks like our lineage and the chimp's diverged around 5 MYA, *after* the split between the human/chimp lineage and that of the gorilla.

*   *   *   *

Before looking at a higher-level example of the same sort of problem, I think we should backtrack a bit and ask about what mental processes are going on in the construction of trees – either correct or incorrect ones. A historical approach is probably the best for illustrating the different kinds of thought process that underlie the decision in favour

FIGURE 24 Two possible patterns of evolutionary relationship between humans, chimps and gorillas. Top: Prevailing view in the mid twentieth century. Bottom: Prevailing view today.

of a particular tree rather than an alternative one. So let's take a brief look back at how tree construction methods have changed over time. The crucial period is the half-century between about 1940 and 1990.

Back in 1940 or thereabouts, how were biologists deciding upon patterns of relatedness? The information at their disposal was largely structural. Plenty of dissections of different creatures had been done, so there was plenty of information on comparative internal anatomy as well as comparative external morphology. Each biologist specialized on a particular group. Knowing lots about that group, in terms of degrees of similarity in various features, allowed the biologist to imagine in what order the evolutionary divergences leading to all the different species of the group had taken place. This 'imagining' was, in a sense, just that. In other words, it was a largely intuitive exercise. No particular, written-down method was employed to construct the tree, although there was a consensus that, in cases of character conflict, some characters – especially embryological ones – were likely to be more reliable than others.

This intuitive method, based on structural information, was used to produce trees for many groups of creatures. The zoological textbooks were full of them. As we will see in a while, some of them turned out to be remarkably accurate and have withstood the onslaughts of later methods and other (molecular) sources of data. Others proved to be wrong.

In the next few decades, two 'methodical' approaches were developed, more or less in parallel. One was the approach called phenetics. This was championed by American biologist Robert Sokal and his British colleague Peter Sneath.[2] The phenetic approach was more quantitative than its intuitive predecessor. It worked by giving each organismic feature, or character, a series of values, or character states. By looking at the degree of 'aggregate similarity' summed across all characters, it was possible to make a tree that reflected varying degrees of overall similarity among the species being compared. For example, in the three-species case, if the aggregate similarity was greatest between horses and dolphins, then, if we wanted to make an evolutionary interpretation of this, we might infer that the lineage separation between these two came later than the split between their common ancestor's lineage and the shark one.

This approach was both better and worse than the intuitive one; better because quantification, however rudimentary, is usually a step in the right direction; but worse because aggregate similarity, at least when used in a simple way, can end up giving important embryological characters and superficial later ones the same weighting. If the intuitive feelings of those 1940ish biologists and their predecessors that embryos are often accurate witnesses to the truth was correct, and in many cases it was, then this egalitarian approach was flawed.

Anyhow, approximately in parallel to the invention of phenetics came the invention of another approach to building trees. This originally went by the name of 'phylogenetic systematics', as I mentioned earlier, but it is more often referred to now by the shorter label of 'cladistics' (from its emphasis on those complete bundles of all the descendants of a single ancestor that we call 'clades'). The cladistic approach was the brainchild of German taxonomist Willi Hennig, who published it first in German, in the 1950s, and then in English[3] in the 1960s.

Phenetics and cladistics had an important difference. While the former employed the criterion of aggregate similarity, as we have seen, the latter did not. Instead, cladistics rejects the use of any characters that are found not only in the group of species whose relationships are under investigation but also in those outside it. These are regarded as primitive characters and discarded. This leaves us to focus on characters that are 'uniquely shared' between any pair of our three species. So, for example, horses and dolphins will uniquely share all those many mammalian characters that sharks lack; dolphins and sharks will uniquely share a few superficial characters; horses and sharks will probably share none, except, of course, for those body-plan features that are common to vertebrates generally.

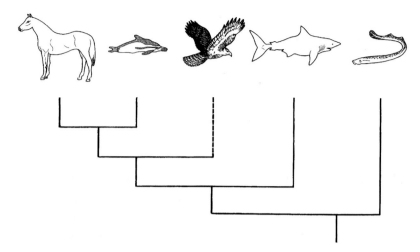

FIGURE 25 The horse–dolphin–shark tree, as shown in the lower picture in Figure 23. This time, two further types of animal have been added: an eagle, which is a disregarded ingroup, and the lamprey (a primitive 'fish'), which is a good choice of outgroup. Jellyfish are also an outgroup, but they are too 'far out' to be useful here.

Now you may have become a little uneasy at this point because mammalian characters are shared by many species outside the horse and the dolphin. But this apparent contradiction is resolved if we define what cladists call 'outgroups'. These are groups that are outside (but close to) the whole frame of reference of the particular study being conducted. They are distinct from 'ignored ingroups' as shown in Figure 25. This solves one problem but it ushers in another. If, before we begin to build our tree, we admit that we don't know the correct pattern of relationships, how do we know which species is a sensible choice of outgroup? Answer: intuition. Back, full circle, to where we started.

After many years of intense debate about which method is best, and indeed which version of which method is best, the philosophical storm has subsided and given way to a sort of hybrid pragmatism. Scientific intuition, quantification of character values, and the use of the outgroup method tend to be combined, in slightly different ways, by different authors. Personally, I think this is fine, though there must still be some purists out there for whom this pragmatism is anathema.

*   *   *   *

The other major change that has taken place in tree building since 1940 is the inclusion of molecular, as well as morphological, data. Protein sequences began to arrive first. The initial trickle became a torrent, and it became a mixed torrent as DNA sequences began to pour into the picture too. This addition of lots of molecular 'characters' was great because the overall information base at our disposal is now much larger, and comparisons based on more information are likely to be more accurate than those based on less, other things being equal. But it did bring with it another problem. This came in the form of a choice. Should we bung all the available data into the computer and ask it to come up with the best possible tree? Or should we put the morphological and molecular data into different files and ask our computer to build two separate trees and then compare them? Computers have slipped in here unannounced; they have become essential in the tree-building process. We humans can handle a few characters in three species without recourse to computers, but if we have (say) 50 characters and 100 species, we are entirely dependent on our lifeless electronic helpers.

Let's now jump back to our starting point and ask why character conflict occurs at all. Surely, if evolution is just gradual divergence and separation of lineages, all characters should tell the same story. Well, if it was, they would; but it is not. As some characters diverge, others converge. And evolutionary convergence is by no means a rarity. If it were, life would be much simpler for tree builders. Convergence in body shape was the cause of the erroneous tree that we started with, in which the dolphin and the shark were considered more closely related than either was to the horse. This kind of convergence is common when there is a double habitat shift – in this case from water to land and then back to water.

No character is immune to convergence, so there is no perfect choice of character to use for building trees. My personal view is that, in general, molecular characters are better. Also, in the morphological realm, I think that the old view that embryonic characters tend to be more reliable than adult ones still has much to recommend it, though we should be careful, as ever, to distinguish embryos from larvae. Sometimes larvae are indeed useful in tree building, but because they are exposed to the outside world while embryos are not, they are perhaps more likely to experience convergent evolution than are embryos.

* * * *

Enough of methods and problems. Let's get back to some actual trees involving actual animals. And, having considered a low-level tree already – the family-level tree involving ourselves, the gorilla and the chimp – let's now turn our attention to a much higher-level tree. This involves the relationships between whole phyla. We are thus dealing with no less an issue than the overall structure of the animal kingdom.

To get to grips with this problem, we need to consider several phyla. But luckily, we don't need to deal with all thirty-five or so of them. I think we can probably get by with just nine. Four of these you are likely to know about already, regardless of whether or not you are a biologist: chordates (including ourselves), arthropods (the biggest group of which is the insects); molluscs (snails, slugs and their marine relatives); and echinoderms (starfish, sea urchins *et al.*). The other five are as follows: tardigrades, hemichordates, nematodes, annelids and nemertines. So the first question is: what manner of animals are they? Tardigrades are tiny arthropod-like creatures whose common name is 'water bears'. The rest are all worms, but don't be fooled by that word into thinking that they are anything more than superficially similar to each other. In fact, they are very distant relatives indeed. Examples of tardigrades and the four different types of worm are shown in Figure 26; and I give very brief descriptions of the worms below to emphasize their differences.

Hemichordates, as their name suggests, are quite close relatives to the chordates. This group includes creatures called the acorn worms (pictured) and also some unwormlike colonial cousins. Nematodes

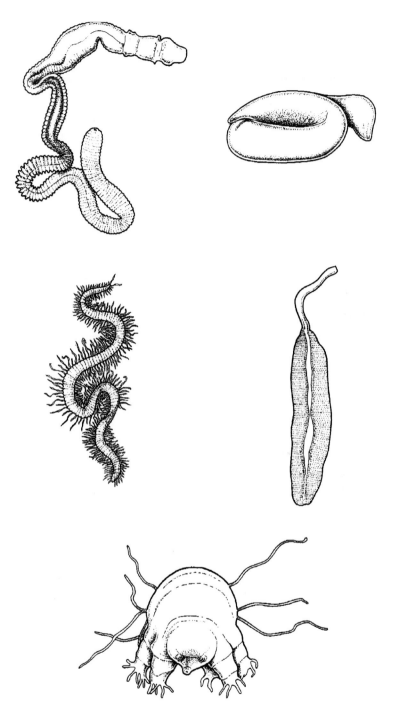

FIGURE 26 Some not-so-well-known animals. Clockwise from bottom: water bear or tardigrade; followed by four very different types of 'worm' (annelid, hemichordate, nematode, nemertine). Sizes range from about 1 mm (tardigrade, nematode) to more than 10 cm (e.g. some annelids).

are called roundworms, because of their simple quasi-cylindrical body form. Annelids are also worms, but they are segmented worms and are generally much larger than nematodes. Finally, nemertines are called ribbon-worms because, although elongate and wormlike, they are generally flat in cross-section rather than round.

What is the pattern of relationship of our overall group of nine phyla? If we can answer this question, we will have given an outline structure to the animal kingdom into which all the other phyla can be interspersed, not that I intend to undertake that latter exercise here.

I show in Figure 27 a view that was popular when I was an undergraduate many years ago, and the view that prevails today. As you can see, there are both similarities and differences, but overall the latter outnumber the former. One 'supergroup' of three phyla – the chordates, hemichordates and echinoderms – remains, though its internal structure has altered so that it might be more logical to rename the acorn worms and their relatives as the 'hemi-echinoderms'. The grouping of the other phyla has changed dramatically. For example, the arthropods used to be grouped together with the annelids in a supergroup called Articulata, whose main common design feature was segmentation. Now instead these two are widely separated in the new tree of life. As a result, either they must have invented segmentation independently (so their segments are not homologous) or, alternatively, segmentation is indeed of ancient origin, and arthropod and annelid segments arc indeed homologous, but in this case other phyla such as nematodes and nemertines must have started off with segments which they subsequently 'lost'. This is a rather unlikely, but not impossible, scenario.

There are several reasons why current views have shifted in favour of the lower tree in the picture. These include the use of a cladistic rather than solely intuitive approach. But perhaps the most important deciding factor was the arrival of molecular data. The initial proposal that phyla should be regrouped in a major way came in 1997, and was based on the analysis of DNA sequences.[4] This looked shaky at first, and many biologists, including myself, were suspicious. But soon, analyses of other molecular data were conducted,[5] and these

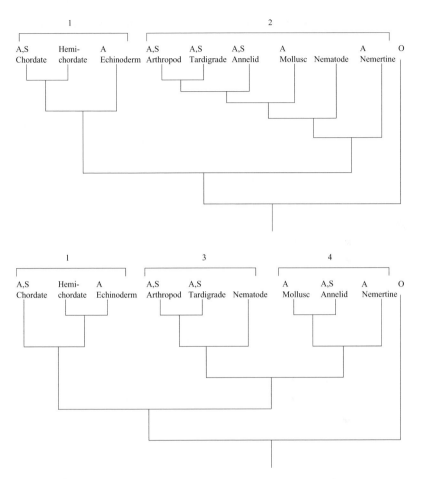

FIGURE 27 An early view of animal phylogeny (top) and the prevailing present-day view (bottom). Names for large-scale (superphyletic) groupings are: 1, Deuterostomia; 2, Protostomia; 3, Ecdysozoa; 4, Lophotrochozoa. 1 + 2 in the top diagram and 1 + 3 + 4 in the bottom one together form the Bilateria, to which O is an outgroup (e.g. jellyfish). A, possess appendages; S, possess segments. (But note that partial segmentation and very minor appendages also characterize some animal groups.)

came up with the same answer. The 'new tree' now seems quite a safe bet.

* * * *

The problem that causes different trees to seem plausible for the same group of taxa is, in this case as in others, character conflict. The use of segmentation pointed one way, while the use of other characters pointed another. And although molecular characters were instrumental in producing the new supergroups, we can see that their constituent phyla do have some morphological features in common. Indeed their names, which I have buried in the figure legend because they are rather offputting, derive from the fact that one supergroup is characterized by a growth process that involves moulting, while the other is characterized by having trocophore larvae. In both cases there are exceptions, but by this time we have learned that evolution is never neat and tidy, so exceptions are hardly a problem.

You'll probably have noticed that I have labelled some of the phyla in Figure 27 with an A, an S or both. This is to represent whether, in general, the phylum concerned has appendages (legs, wings, fins, etc.) and/or segments (whether externally obvious or not). The reason for this is that in the next chapter I will focus on two of the classic case studies of comparative developmental genetics. These are based on particular genes that are involved in the development of segments and appendages. These are the 'stripes' and 'spots' respectively of the chapter title. So now, armed with some understanding of the world of trees, we enter the world of genes, with a view to connecting these worlds, and that of the organism, together.

15 Stripes and spots

If we judge the advent of evo-devo in the 1980s from the standpoint of evolutionary modes, it conforms better to the 'explosion' mode than to the 'plodding' one. The spark that lit the fuse was the discovery of this thing called the homeobox. So I had better explain that first. It emerged, as many scientific discoveries do, in two separate places at about the same time – a bit like Darwin and Wallace coming up, independently, with the idea of natural selection. In the case of the homeobox, the two groups that made the discovery were in Indiana[1] and Switzerland.[2]

The homeobox is a stretch of DNA that has a particular sequence of building blocks (nitrogenous bases). In total it has about 180 of them. This is a short stretch of DNA when compared with the total length of a typical gene, which has thousands of bases. So when biologists use the phrase 'homeobox gene', they do not mean that some genes *are* homeoboxes – rather that they *have* homeoboxes. That is, somewhere along the length of the whole gene there is a homeobox; and exactly where this 'box' appears varies from gene to gene.

As you may already know, geneticists have a habit of calling various recurring sequences of DNA 'boxes'. This particular box gets its 'homeo' prefix from the fact that it was initially discovered in genes that caused particular kinds of mutant phenotype, where the right thing appears in the wrong place – like the 'antennapedia' fly that I mentioned earlier with legs growing out of its head. Since these 'right-thing-in-wrong-place' mutations are called homeotic mutations, it made sense to label a piece of DNA that was found in such genes the 'homeobox'.

But what became clear as the molecular work progressed was that homeoboxes were not just found in these genes. They were found in others too. And, amazingly, they also began to show up in lots of different creatures, including some in which homeotic mutations were unknown. The animals that were first investigated ranged over several phyla, and since then many more animal phyla, and several groups of plants, have been examined and also found to have genes with homeoboxes. It is now clear that *all* animals and plants have homeobox genes; but the very simplest of creatures – bacteria – lack them.

Finding a DNA sequence that was highly conserved across such distantly related organisms whose evolutionary lineages had been separated for hundreds of millions of years was unexpected. It looked like homeobox genes were doing some fundamental job that was indispensable. That idea turned out to be true, and the indispensable job turned out to be the switching on or off of other genes. This, as you will recall, is crucial to developmental cascades, where 'upstream' genes have switching effects on 'downstream' ones.

This raises the question of exactly how one gene switches on or off another. In cases of direct control, the controlling gene's product needs to bind to the regulatory region of the target gene. So this protein product must contain a particular stretch of amino acids that fulfils this binding role. You've probably guessed it by now. Those highly conserved homeobox regions code for the parts of the protein products – called homeodomains – that bind to the DNA of the target genes. So possession of a homeobox identifies a gene as being involved in the control of others, and this generally means that homeobox genes are developmental genes. However, not all developmental genes have homeoboxes, because some operate in different ways.

The two examples of developmental genes that I'm going to deal with in this chapter are both homeobox genes. Their names are *engrailed* and *distal-less*. We met *engrailed* in Chapter 4, where we noted that it is involved in the formation of segments (though it does other things too that I'll ignore here). The most famous role of *distal-less* is in the development of limbs, though it, too, has other roles, and I will be including a brief look at one of these (the development of the butterfly eyespots that we discussed in Chapter 12).

FIGURE 28 Stripes and spots of gene expression. Top: Stripes of expression of the gene *engrailed* in a centipede embryo. Bottom: Spots of expression (growing into cones) of the gene *distal-less* in a developing insect. (Anterior is to the left in both cases.)

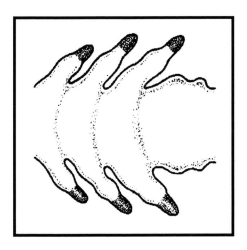

The current chapter's title derives from the fact that these two genes are often expressed in spatial patterns (Figure 28) that can be described as stripes (*engrailed*) and spots (*distal-less*). When you think about it, this makes sense. Segments are essentially transverse subdivisions of the body. These are most obvious in animals that are clearly segmented from an external observer's viewpoint, such as the arthropods. Limbs, on the other hand, grow out from particular points (or 'spots') of the trunk. So it makes sense that genes that are involved in the formation of segments should be expressed in transverse stripes, whereas those that help to initiate the outgrowth of limbs should be expressed as spots.

I'd better sound a note of caution here that applies to both of my examples, before proceeding to deal with each separately. My story here is a simplified one. While this is true of the whole book (and indeed of all books that attempt to get an area of science across to a wide readership), it is particularly true in the present chapter. The field of comparative developmental genetics has revealed many things. One of these is that the involvement of genes in the developmental process of any organism is exceedingly complex. So I am very conscious of the wealth of detail that I am omitting.

There are two aspects of this. First, as I already noted, both of the genes that I will be focusing on have other developmental roles than those that I will be discussing. This is now known to be a widespread feature of developmental genes generally. Second, the developmental cascades for segment formation and limb formation involve many genes in addition to the ones that I have selected for discussion. Some of these other genes are expressed upstream in the relevant cascades, some downstream. That is, *engrailed* is neither the first nor the last gene to be expressed in the long sequence of events leading from the first broad subdivisions of the anteroposterior axis to the finalization of the segments in all their splendid morphological detail. Nor is *distal-less* the first or last player in the limb development story. But in a way, none of this matters. We can discover interesting, even awesome, things from the simplified version. So let's now get into the two stories – stripes first, then spots.

* * * *

We'll start by going back to the dawn of animals, in the almost impenetrable darkness (from a fossil viewpoint) of pre-Cambrian times. One thing about early animal evolution seems clear, concerning the symmetry of the body. The first animals were asymmetric. This type of body plan is still represented in today's fauna by sponges. When asymmetry gave way, in some lineages, to a form of symmetry, what emerged first was radial symmetry. Today's jellyfish retain this design. But, even before the beginning of the Cambrian, some animals had become bilaterally symmetrical. That is, they had head and tail ends, left and right sides. Perhaps the first animals to adopt this kind of body plan were primitive flatworms, a bit like the group still around today called 'acoel flatworms'; but that's just one hypothesis out of many.

Most of today's animals are descendants of these first 'bilaterians'. Almost all the groups that I have talked about at various stages have a head, including vertebrates, arthropods, molluscs and so on. The main exception is the group called echinoderms, which, as you'll recall, includes starfish and sea urchins. But these are what is called 'secondarily modified'. That is, their ancestors were once recognizable bilaterians even though the present forms are not.

It is only in the realm of the bilaterians that the concept of segmentation, as we normally use it, can be applied. This is because until there is a head end and a tail end there is no anteroposterior axis. And 'segmentation' means the division of the animal concerned into discernible subunits along this particular axis. So a fly, for example, has a head, which is a fusion of several segments, a thorax, which has three segments, and an abdomen, which has about nine segments. The head, thorax and abdomen are not themselves referred to as segments; rather, because they are higher-level units, they are given a different name – 'tagmata' ('tagma' in the singular).

The way in which segments are formed was first worked out in flies, and is beautifully described by the British biologist Peter Lawrence in his book[3] *The Making of a Fly*. I dealt with this, in my usual simplified way, in Chapter 4. It is apparent from that account, and in more detail from Lawrence's, that there is a segmentation cascade and that *engrailed* is a major player in this. Although it is neither at the beginning nor at the end of the overall cascade, it is an important milestone along the way, because it is a member of a group of genes (called the segment polarity group) whose expression pattern takes the form of 'one stripe per segment'. It is therefore one of the first 'segmental markers' from which you can determine the number of segments that will be visible to the naked eye later on.

Now we have a choice that you might describe as a vertical versus horizontal one. Either we can go deeper down into more molecular depth for flies (vertical) or we can broaden our horizons to look at the way in which segments are formed in other animals (horizontal). It will come as no surprise to you that I will now head off in the latter direction. We are doing *comparative* developmental genetics here; that, after all, is the route to understanding more about how development evolves.

So the first question is: which animal phyla are characterized by possession of a segmental body plan? This is not such an easy question to answer as you might think, because some phyla exhibit what might be described as partial segmentation – for example segmentation of the external cuticle but not of internal body parts. I will ignore these here, and concentrate on two batches of phyla: those that clearly *are* segmented (chordates, arthropods and annelids) and those that, despite being bilaterians, are not (like the nematodes). This latter, unsegmented group actually comprises the bulk of the animal kingdom in terms of numbers of phyla, even though, in terms of known species numbers, they are all eclipsed by the biosphere-dominating arthropods.

Figure 29 is a diagrammatic evolutionary tree showing the pattern of splitting of lineages that led to the 'big three' segmented phyla. It also shows three possible scenarios for the evolutionary origin of segmentation. As can be seen, these are the 'single origin', double origin' and 'treble origin' scenarios.[4] The big question is: which of these is correct? Was segmentation of the body invented once, twice, or three times in phylogenetic history? And, before we can answer these questions: what kind of evidence is relevant here? What facts should we bring to bear on this difficult issue in the hope of resolving it?

First of all, we need to be fairly confident that our overall view of the relationships between animal phyla is correct. That is, that the lower panel in Figure 27 shows the pattern of lineage splitting that actually took place. Prior to the late 1990s, a different pattern of relationships (upper panel) was generally agreed upon, as noted earlier, in which the arthropods and the annelids were sister groups. So they were thought to be very closely related to each other.

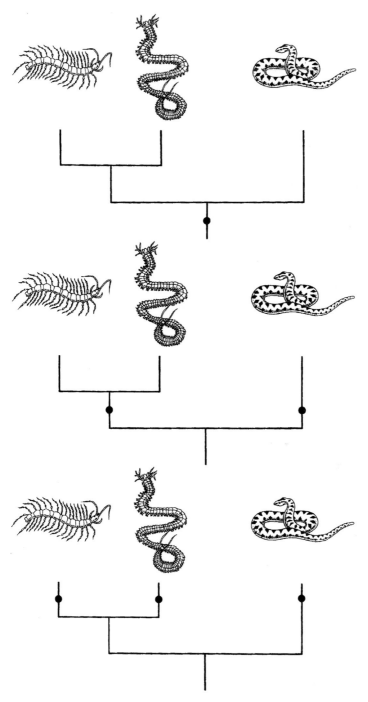

FIGURE 29 Three possible scenarios for the evolutionary origin(s) of
animal segmentation. Solid circle, origin. From top to bottom: single-,
double-, treble-origin scenarios.

This picture of animal phylogeny went hand in hand with support for the double-origin model of the appearance of segmentation, which was the prevalent view for many decades. An account of it can be found in the 1964 book[5] *Dynamics in Metazoan Evolution: The Origin of the Coelom and Segments*, written by English zoologist Bob Clark. If the current view of the animal kingdom is correct, with arthropods and annelids being distant rather than close relatives, then the argument in favour of a double origin becomes much weaker. These days, most biologists seem to favour one of the other models: either a single origin or three.

* * * *

Many scientists have an almost innate preference for simple answers. Using the simplicity/complexity of alternative hypotheses as a guide to which is more probably correct is referred to as using the principle of 'parsimony'. Whichever hypothesis is the simpler (that is, the more parsimonious) is to be preferred, other things being equal.

Which of the one-origin and three-origins hypotheses is simpler? Regrettably, the answer depends on what criterion you use to measure simplicity. On the face of it, postulating that a complicated evolutionary event, such as the invention of segments, occurred only once seems inherently more parsimonious than postulating three independent origins. However, the latter view seems more parsimonious when we consider also all those unsegmented bilaterian phyla. The point is that if segmentation arose only once in the history of the animal kingdom, it must have arisen very early, before what is called the 'radiation' of the bilaterian phyla. But if so, then we would expect all these phyla to have segmented body plans, and this expectation is not borne out. Indeed, the vast majority of bilaterian phyla are unsegmented.

Such a pattern is not entirely inexplicable on a single-early-origin hypothesis. After all, many things that arise in evolution later disappear. The loss of eyes in the cave-dwelling descendants of open-environment ancestors is a favourite example of this. If eyes can be lost, so can segments. But the number of these losses must be very large under the single-origin model, while if arthropods, annelids and chordates all invented segments independently, then the number of losses that it is necessary to postulate becomes much lower, and could even be zero. So the hypothesis that is less parsimonious in terms of the number of origins is more parsimonious in terms of the number of losses. Stalemate.

To summarize so far: the current view of phylogeny reduces the likelihood of the double-origin model, but considerations of parsimony do not give us a clear pointer towards either of the other two. It is thus time to bring in a new type of evidence. So now we return to where we started in this chapter – comparative developmental genetics. But what is the underlying rationale here? How will comparative information about the ways in which genes are involved in segment formation in the various groups of segmented animals help to answer our question about the evolutionary origin(s) of segmentation?

To answer these questions, we need to go back to the concept of homology. Let's for the moment forget segments and grand-scale, interphylum comparisons, and consider something simpler in order to refine our understanding of what homology means and what the alternative to it is. Compare, for example, your left arm and a mouse's left foreleg. These are homologous structures, as can be seen from the similar arrangement of constituent parts, such as the bones. But homology often goes deeper than just resemblance of adults. Given that our arms and the forelegs of mice were derived from a common forelimb in a distant mammalian ancestor, and given that their evolutionary modification in different directions was based on mutations that reprogram the developmental trajectory of the forelimb bud at various stages of its ontogeny, it would be reasonable to expect to see similarities in forelimb *development* too. And these we do indeed see.

Similarities in the development of different mammalian limbs were known long before the era of developmental genetics. Now that we are in this era, it is apparent that as well as the morphological similarities of embryogenesis, there are genetic similarities. To a large extent, the same genes are involved in the development of human and murine limbs. Not exactly the same genes, of course, because genes, like limbs, evolve over time. But genes that are so similar that they in turn look like homologues of each other. So homologous developmental genes are involved in the production of homologous structures.

Homology is used to describe situations in which certain body parts (limbs, genes or whatever) of different creatures have partial similarity (and therefore also partial difference); and the similarity is due to the inheritance with modification of the parts concerned from a common ancestor that already had some version of them. The alternative explanation of any situation where we find both similarities and differences is that the parts concerned were not present in the last common ancestor of the two animals we are comparing, but rather have been invented independently in the two lineages at some point since their divergence. In this case, the similarities are due to shared selective pressures rather than shared ancestry. This kind of evolutionary process, where different lineages are pushed into coming up with at least superficially similar solutions to the same problem, is called convergence. Indeed, we came across this before when considering the problem of 'character conflict' affecting the building of evolutionary trees in the previous chapter.

A good example of convergence concerns the shells of bivalve molluscs (cockles, mussels, clams, oysters, etc.) and the shells of a less well-known group of animals called brachiopods (a separate phylum from the molluscs altogether). It would seem reasonable to go along with the argument that a soft, highly edible invertebrate living in a sea full of predators would do well to protect itself with some hard outer casing or shell. This can be done in various ways. One group of molluscs (snails) devised a single shell; another, the bivalves, devised the different double-shell arrangement that their name implies. But the brachiopods, which to an inexperienced observer would simply be mistaken for bivalves, also devised a double-shell design. There are, as before, both similarities and differences. This time, although the external appearance is very similar, examination of the internal anatomy reveals that whereas in bivalves the two shells are essentially left and right plates of armour, those of the brachiopod are upper and lower ones.

Usually, there are 'giveaway signs' like this to tell us that an observed partial similarity is not a homology but rather a result of convergence. The contrast between the horizontal whale's tail and the vertical fish's tail is another good example. Anyhow, in cases where convergence turns out to be the correct explanation, the expectation is that the ways in which the partially similar adult characters develop will be quite distinct. And that is the case with bivalves and brachiopods, though it has to be said that not a whole lot is known about brachiopod embryology.

So, we might expect that homologous characters would be produced through similar embryological processes that in turn are underlain by developmental cascades involving similar groups of genes. Equally, we might expect that convergent characters would be produced by different embryological processes underlain by developmental cascades involving different groups of genes. This contrast in expectations is fine, but sometimes it is not borne out by observation. Sometimes, evolution extensively modifies the development of homologous characters in one or both of a pair of diverging lineages; and sometimes, converging lineages accidentally hit upon similar ways of developing the characters concerned, as we will shortly see.

Despite this potential complication, which will cause a certain error rate, it is often possible to reach the correct decision about whether partially similar characters in a pair of species are homologous or convergent by looking at the degree of similarity of their development at either descriptive or genetic levels.

*   *   *   *

Now, back to segmentation. Before we proceed to examine, at the genetic level, the degree of similarity of the developmental processes leading to segmentation in arthropods, annelids and chordates, I should nail my colours to the mast. I support the three-origins model. I do not believe, as some do, that the very first bilaterian animals were segmented. Of course, I may be wrong. It's early days yet in this fascinating controversy. But if I and other advocates of the three-origins model are right, then the expectation is that the developmental genes involved in the production of segments will be very different in the three phyla concerned. This is simply because, with genomes consisting of many thousands of genes in each phylum, the chances of hitting on the same ones to produce segments should be very low if they are truly independent 'inventions'. However, this expectation is not borne out. Many homologous genes are involved in the development of annelid, arthropod and chordate segments. Here, I'll restrict myself to the *engrailed* story.

The one-stripe-per-segment expression pattern of *engrailed* that was first discovered in the fruitfly has now turned up all over the place. Not surprisingly, it is found in other insects. It is also found in the other classes of arthropods. Since most biologists are now agreed that the arthropods are a natural group that radiated from a single ancestor, this finding ought not to be too surprising either. But it is now clear that a broadly similar expression pattern of *engrailed* homologues occurs in at least some annelids (leeches) and at least some chordates (the protochordate called amphioxus and the zebrafish).

There are two reasons why I have not abandoned my support for the three-origins model of segmentation in the face of such apparently contrary evidence. First, the expression patterns are indeed only *broadly* similar. For example, in the zebrafish the segmental stripes of *engrailed* expression show up later than their arthropod counterparts, so they may play a somewhat different role in the process of producing segments. Second, there is a reason why the odds against independent evolutionary inventions of segmentation hitting upon the 'same' developmental genes may not be as low as they first seemed on consideration of the awesome size of genomes. In fact, they may be quite high.

We have now arrived at something called co-option, which I believe to be one of the most important ideas in the whole of evo-devo. I can explain it best by a 'let's suppose' scenario. Let's suppose, then, that the first ever bilaterian animal was indeed some sort of unsegmented flatworm. But suppose that it has some form of repetition of internal structures, such as nerve ganglia. After all, most animals do. There must be a cascade of developmental genes overseeing this repetition. Like all other developmental cascades, it is likely to be complex, involving many genes. If, much later in evolutionary time, any of the many descendant lineages become segmented, what is the 'easiest' way to achieve this? That is, what is evolution, from a probabilistic standpoint, most likely to do? One possible answer to this question is that it is far more likely to 'co-opt' some existing cascade of genes by expressing those genes in parts of the embryo where they had not been expressed before than to build up, *de novo*, some entirely new cascade. If there is any truth in this argument, then convergent structures might be expected to be underlain by partially similar systems of developmental genes rather than entirely different ones.

* * * *

It's now time to turn from segments to limbs (or appendages) and see if we encounter a similar story. But before getting into the details, it is worth reflecting for a moment on what exactly is meant by a limb. Although most of us tend to picture a leg when the word 'limb' is mentioned, the term can be used very widely for just about anything that 'sticks out'. So, in the animal kingdom, there are many types of limb. In addition to legs there are wings, arms, fins, tentacles, antennae and various forms of sticking-out mouthparts. Whereas unsegmented phyla are more numerous than segmented ones, phyla in which some sort of limbs are found probably outnumber those in which there are no limbs at all. Some body plans that we tend to think of as limbless do indeed have limbs in at least some of their constituent species. For example, the term 'worm' conjures up a featureless cylindrical body form with no limbs at all. Yet ribbon worms (nemertines; Figure 26) have a long protruding mouthpart (the proboscis) and some annelid worms have lots of lateral 'parapodia'; this is true of many marine annelids, though not, of course, of earthworms.

Like segments, limbs are quasi-autonomous developmental modules, as I argued in Chapter 13. Such modules are caused to develop in their characteristic ways by the appropriate cascades of interacting developmental genes. As usual, the first breakthrough came with work on the fruitfly, which is the model system in which *distal-less* was first discovered. This gene is switched on at very precise places – 'spots' – in the thoracic segments.[6] The trigger for a spot of expression seems to be the combination of upstream gene expression that occurs at the three-way intersection point between anterodorsal, anteroventral and posterior compartments. The expression of *distal-less* marks the point from which outgrowth of the leg will occur at metamorphosis. (Bear in mind the fact that fly larvae are legless, though some other insect larvae do have legs of sorts, so leg formation, from the appropriate imaginal discs, occurs much later than embryonic development.)

We broaden our horizons now, just as we did when looking at segmentation. We extend from our fruitfly base to other animals with limbs. I'll be selective here, partly because it is my usual *modus operandi*, but partly also because there are a lot of obscure types of limb in little-known phyla where the genetic basis of their development is as yet entirely unknown. Let's focus this time on just two types of animal – insects and starfish. These two, as we are already aware, are extremely distant relatives. Starfish have limbs of an unusual kind – the tiny 'tube feet' that cover the ventral surfaces of what are ironically called 'arms'. I suppose these are also limbs at a higher level. So here we have a hierarchical limbs-within-limbs situation. Anyhow, it is the tube feet that I want to focus on. There are hundreds, perhaps thousands of these, in contrast to the small number of arms (usually, but not always, five).

If we use the same techniques to look for places where *distal-less* is switched on in starfish as were used to look for *distal-less* expression in the fruitfly (and I don't intend to go into the methodological details here), what do we find? I suspect that you have already guessed the answer. There is a tiny patch of *distal-less* expression at the end of each developing tube foot. This is also true of many other types of limbs in many other animals[7]. This consistency of expression pattern makes *distal-less* look like a gene that is good at producing a particular kind of topography – outward projections. But I noted earlier that it was also involved in the development of butterfly wing eyespots. Here, although there is no outgrowth – the wing is just as flat in the eyespot regions as elsewhere – there is nevertheless still something in common between eyespot and limb expression of this gene. In both cases, the expression takes the form of a spot rather than a stripe.

It would be hard to find a biologist who thinks that the legs of a fruitfly and the tube feet of a starfish are homologous. It seems that here, as with segments, we have found evidence for the co-option of cascades of developmental genes for new roles. Just as the ancestral bilaterian animal may have lacked segments but possessed repeated internal structures like nerve ganglia, it may have lacked legs but possessed some rudimentary form of outgrowth, even perhaps an 'internal outgrowth', which is not the contradiction in terms that it sounds – the appendix in your gut is just such an internal outgrowth. If so, then the 'easiest' way to make more elaborate outgrowths of different kinds in the various descendant lineages may well have been to make use of an existing cascade by expressing it in different places. That would seem to have a higher probability of occurrence than the appearance of a whole new suite of interacting developmental genes.

*   *   *   *

So, two genes, same story. The punchline in both – co-option. The idea that this process is of general importance in the evolution of development is the main message of this chapter. There are lots of complications I have glossed over, but I don't believe that they will end up detracting from the general message. I'll mention just one of them, namely gene duplication. This, in some other contexts, is a general message in itself, as we saw in Chapter 13. However, in the present context it can be problematic in terms of being sure that we are making the right comparison.

What I mean is this. Sometimes, when we want to compare two types of animal, the lineage leading to one of them has experienced a duplication of the gene that is our focus of study. So the animal at the end of that lineage has two copies of the gene, whereas the animal at the end of the other lineage still has just one. But gene duplication is frequently followed by divergence of function, and that will in turn involve divergence of expression pattern. So which of the duplicated genes in one lineage do we compare with the single gene in the other? This is often a difficult question to answer. For now, it is probably best to stick to making comparisons between lineages where each has a single copy of the gene concerned.

For the longer term, though, this approach is inadequate. Evolution is not a collection of independent processes. Everything interacts with everything else, albeit to varying degrees. So we need to connect things up. With regard to the connection between duplication and co-option, perhaps the latter is more likely to occur following the former. And perhaps when a group of genes that interact functionally are also close neighbours on a single chromosome, a large-scale duplication event could be the starting point for the divergence and co-option of a whole new cascade (or of a side branch of the existing one). This sort of process appears to have occurred many times in the evolution of the Hox genes.

16 Towards 'the inclusive synthesis'

I like to think of this book as a contribution to a growing movement whose goal is to transform the 'modern' synthesis of the twentieth century into a more inclusive synthesis that is appropriate for the twenty-first. And, having looked at the various pieces of the puzzle over the last several chapters, we are now in a position to try to put them together. I don't want to overstate what can be achieved at the present time, when so much research in evo-devo still remains to be done. That's why I put 'towards' in the chapter title. But I don't want to understate the case either. We can already build a significantly more inclusive synthesis, even if further inclusivity is yet to follow.

There are two possible ways in which the old synthesis might be affected as it evolves into the new, or, perhaps more accurately, a spectrum of possibilities between two extremes. At one end of this spectrum lies the possibility that the old is unaffected by the new, in the same way that a building may be largely unaffected by an extension built on to one of its sides. At the other end lies the possibility that the old is effectively demolished by the new, as when a crumbling old building is bulldozed to make way for a new one (Figure 30).

My current view is that the relationship between the 'modern synthesis' and the 'inclusive synthesis' is somewhere in between these two extremes. I believe that much of the integration of ecology, genetics, palaeontology and systematics that was achieved in the twentieth century will not just survive, but will be – indeed is being – enriched by the addition of new information on development. However, I also believe that the more extreme pronouncements of *some* neo-Darwinians, particularly in relation to the factors that determine the direction of evolutionary change, need to be bulldozed and

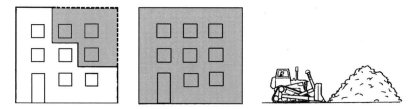

FIGURE 30 Two possible relationships between new and old actual or theoretical edifices: extension versus replacement.

replaced by more pluralist views. I like to consider this as a return to the true spirit of Darwinism.

I have already criticized one of the architects of the synthesis in this respect (Fisher). But Fisher's work is now receding into the historical past, his main work[1] having been published in 1930. What of more recent neo-Darwinian views? Might it be that everyone is now happy with the kind of role that I am proposing here for developmental bias as a determinant of evolutionary directionality? Might there be no one left to fight but strawmen?

Let me address this important issue by examining the way Richard Dawkins, perhaps neo-Darwinism's leading present-day exponent, at least in terms of the more 'popular' literature, began his highly successful book[2] *The Blind Watchmaker*. Here are his lead-in words: 'This book is written in the conviction that our own existence once presented the greatest of all mysteries, but that it is a mystery no longer because it is solved. Darwin and Wallace solved it, though we shall continue to add footnotes to their solution for a while yet.'

Is the whole new evo-devo endeavour a footnote? I think not. Is the proposal that developmental bias plays a major role in determining the direction of evolutionary change a footnote? Hardly. We are entering an exciting new era of evolutionary theory in which development is at last beginning to make the contribution that it must. To write this new era off as a footnote (and it had already begun when Dawkins wrote those words) is in my view a serious mistake. I would agree that Darwin and Wallace largely solved the external side of the evolutionary puzzle, but the internal side is only now yielding some of its many secrets. And note that it is a 'side', not a 'footnote'; or, if you prefer, an equal partner rather than a bit-part player.

Given that we have arrived back again at the internal/external contrast (or 'metaphor'[3]), it may be profitable to analyse its nature a bit further before proceeding. This is especially wise given that it is a multifaceted contrast rather than a simple, singular one.

Pick up an introductory undergraduate text on evolution at random. If it is a fairly recent one it will probably include at least a token chapter on evo-devo. But of course, it will also include accounts of all those famous old stories of adaptation such as Darwin's finches. What you will probably find is that the evo-devo material is organized around homeobox genes and their developmental effects, with some speculation about their evolutionary significance. The comparative element will probably be at a high taxonomic level, like arthropods versus chordates. In contrast, the material on adaptation will probably focus on the environment of the organisms concerned, a description of the relevant ecological niches, and a discussion of the ways in which externally observable phenotypic characters, usually in the adult and of unspecified developmental origin, are such that they can be seen to contribute to the adaptedness of the creatures concerned. The comparative element in this case is usually small-scale, such as across several congeneric species.

Personally, I find this kind of split approach unhelpful, even misleading. It engenders a view that there are two kinds of evolutionary process going on – a macroevolutionary evo-devo process and a microevolutionary neo-Darwinian one. This, of course, is nonsense. The authors of such accounts can be excused up to a point, because we still don't know much about the developmental genetics of variation in beak size or about how important developmental genes, such as those with homeoboxes, evolve in the wild at the population level. Nevertheless, lack of some bits of information should not be an excuse for lack of clear thinking.

The risk that we need to avoid is that of confusing the following: the distinction between adult phenotypic characters and their developmental origins; the distinction between adaptation to the external environment and internal coadaptation of different parts of the organism; the distinction between developmental bias and natural selection as evolutionary drivers; and the distinction between large-scale and small-scale evolutionary processes. Contrary to the restricted 'textbook' view that I outlined above of the relationships among these four distinctions, everything is mixed up together. So great care is needed in any discussions of internal versus external factors in evolution. There is not just one kind of internal versus external contrast; there are several, and they don't align in a simple way.

* * * *

What I'm going to do now is to run through each of the seven major messages of the book, in the order that we first encountered them. I'll try to summarize each and also to explore their interconnections. Then I'll come back to the issue of the degree of compatibility of the 'modern' and 'inclusive' syntheses.

1. The return of the organism. For me, one of the great achievements of evo-devo has been putting the organism back in a centre-stage position in our quest for an inclusive theory of evolution. Yes, populations are important, for all evolution happens in them. And yes, so are genes, for these are the bearers of the codes for making organisms from one generation to the next. But the way the organism got squeezed out of the picture by Fisher's abstract algebra of the changing gene frequencies in evolving populations was a tragedy. The culmination of this, for me, was the definition of evolution as a change in the gene frequency of a population. There is a link to 'message 2' here because if you define evolution in such a restrictive way, it naturally leads to the downplaying of the *introduction* of variation as opposed to its subsequent sorting.

The lack of focus at the organismic level that characterized much of the modern synthesis robbed us, until recently, even of the basic language to use to describe evolutionary changes at this level. Genes change by mutation; populations change by selection; organisms change by . . . well, we can now call it developmental reprogramming. Not just heterochrony, nor even just that plus its spatial counterpart heterotopy, but *all* possible changes in the developmental process including those two. Reprogramming is a more complex process than mutation, because it is a change in something that is itself by definition a state of change. Development is a trajectory through multicharacter hyperspace; developmental reprogramming is a mutationally induced change in that trajectory. But a new phrase is the beginning, not the end, of our quest.

2. *Developmental bias and 'possible creatures'*. This approach leads to a very different view of the determinants of evolutionary direction from the alternative 'actual creatures' approach, especially in the latter's most extreme form of 'evolution as changing gene frequency'. The main message here is that the structure of the variation upon which natural selection acts is an important player in the evolutionary game. This is true regardless of whether we are dealing with populations that have considerable 'standing variation' or those that are 'waiting for mutation'.

The idea of the organism as a piece of putty that can be moulded in any desired direction by all-powerful selection is too simple. Rather, biases in mutation, in the ways in which development is reprogrammed, and in the structure of the standing variation are all potentially important causes of the direction that evolution takes. They do not oppose selection; rather they interact with it, and it is this interaction that sets the evolutionary sails. This is perhaps the prime area for bulldozing the old (at least in its extreme Wallace–Fisher–Ford–Dawkins manifestation) and replacing it with the new, broader, view. I believe that this is the most important of my seven 'major messages', which is why I (a) named the book after it, and (b) will focus almost exclusively on it in the final chapter.

3. *Internal selection and coadaptation*. This is an area where I think that in the end there is no real problem of compatibility between the 'modern' and 'inclusive' views. There has *seemed* to be such an incompatibility, but I don't believe that it is real. I think this is another language problem, but this time not the absence of a necessary term

but rather the presence of an unhelpful one. The culprit, in my opinion, is the term 'internal selection'. Many have refused to accept this, a recent example being the Italian biologist Giuseppe Fusco.[4] Inasmuch as such objections are centred on the term rather than the concept, I agree with them, because I have come around to the view that 'internal selection' is too open to misinterpretation to be a sensible choice of phrase. If, instead, we talk about selection for coadapation, few will seriously object. And if we acknowledge the continuum from pure ecological adaptation to pure internal coadaptation that is embodied in the idea of a trans-environment fitness profile (Chapter 10), then I don't believe that there is anything much to argue about here. Instead of argument, what is needed is action, in the form of case studies of coadaptation that are good enough to rival the best of their ecological–adaptation counterparts.

4. The developmental reaction norm. This, you will recall, is the phrase used to describe the influence of the environment on the developmental process. It can be looked at in two ways. One view is that it is nothing terribly new or important because quantitative geneticists have known for approximately a century that most phenotypic characters are only partially heritable, and that the environment, as well as the genome, plays a role in determining the values that these characters take.

But a counter-view goes something like this. Much evolutionary theorizing in the neo-Darwinian tradition has been as neglectful of quantitative genetics as it has of developmental biology. Many adaptive scenarios have been modelled in which a character's value is determined by the genotype; and the environment only comes into play after the character values have all been set. So, selection sieves these values, letting only some pass through into the next generation (or, in less drastic form, giving them different probabilities of 'getting through').

If the environment helps to determine the value of the character, it has a dual role that many evolutionary biologists have chosen to ignore. And if some genes render the effect of the environment on development greater or less than (or just different from) other genes, then selection can alter their frequencies and in doing so can alter the way in which the genome and the environment interact to influence the course of development. This is a very different, and much more complete, picture of the environment's role in evolution from the 'consider a genotype AA that produces character value X' type of approach. Although the degree of difference between the two approaches varies from one character to another, few developmental characters can be said to be entirely unaltered by any form of environmental variation.

5. *The existence of quasi-autonomous modules.* This is a very topical subject, and one to which whole conferences are devoted. It is an old concept that has been given a new lease of life by the advent of evo-devo. It links in with the idea of evolution by duplication and divergence, which is important in both the molecular and morphological realms. The best way to treat duplication and divergence within my overall conceptual framework is as an important and general form of developmental reprogramming (message 1). Of course, reprogramming takes many other forms too (like heterochrony). But, given that this is true, an interesting question arises: can we predict under what circumstances evolution will incorporate some of these general types of reprogramming and under what (different) circumstances it will incorporate others? The 'circumstances' include genetic, developmental and ecological ones. There is a huge research programme lurking here, and just beginning to come into view.

6. *Phylogeny reconstruction.* Although the 'old' synthesis included palaeontology and systematics, these became less conspicuous over time in what has been described as the 'hardening' of the synthesis. I suppose it is true to say that they were absent at the beginning, present in the middle, and almost absent again at the end of the period concerned (say from 1930 to 1970). Fisher's theoretical beginnings paid scant attention to phylogeny; Mayr and Simpson rectified this omission as the synthesis was 'fleshed out' in the 1940s and 1950s; then Ford abandoned it in the almost history-free ecological

genetics approach of the 1960s. None of these decades was as homogeneous as that simplified story suggests; but in terms of the *relative* emphasis on phylogenetic matters, there is some truth in my generalization.

While some sorts of general evolutionary principle, including natural selection, are applicable regardless of particular patterns of relatedness, other general principles are not. If we seek generalizations about the way in which the different forms of developmental reprogramming are involved in evolution, then phylogeny matters. If we believe that an animal with similar segments represents a primitive state and one with different segments (e.g. organized into tagmata like head, thorax and abdomen) is more 'advanced' or derived, then we may treat this as a story of the divergence of replicated parts. If we are wrong, and we have mistaken the primitive for the derived, and vice versa, then an 'opposite' form of reprogramming has taken place. Both, presumably, could be driven by natural selection, but only one of them has been. And it would be nice to know which. So a non-phylogenetic approach, while acceptable to some (and only some) population geneticists, is unacceptable to students of evo-devo.

7. *The concept of co-option.* This, at least in relation to the genes whose products interact in developmental cascades, is genuinely new – a child of evo-devo research. Like modularity, it falls under the banner of general types of developmental reprogramming, but now at the molecular level. Although it might appear that there is no conflict between the concept of co-option and the 'modern synthesis', because the synthesis pre-dated comparative developmental genetics and so was unable to make any contribution in this area, this may turn out to be a misguided view. Co-option relates to the relative ease of modifying embryos in different ways, and so links up with my main message of developmental bias and its role as an evolutionary driver. Furthermore, there are probably many different types of co-option, the differences between which are just beginning to emerge. Some may be more important than others. One particularly important form may be the simultaneous co-option of a whole 'cassette' of

interacting developmental genes; this may be crucial in the origin of evolutionary novelties such as limbs.[5]

*   *   *   *

I hope that the above messages have persuaded you, if you were in need of persuasion, to agree with the view that I stated at the beginning of this chapter, about the relationship between the 'modern' and 'inclusive' syntheses. That is, the view that this relationship is somewhere in between the extremes of 'bolt-on extra requiring no modification of the original' and 'better version requiring demolition of the original'. Much of the modern synthesis is a population-level endeavour that is being enriched by the organism-level findings that are pouring out of the evo-devo camp. However, a few bits of it will be replaced rather than enriched. Not everyone will agree with this view. Some will say that I have been too kind to the modern synthesis, some that I have been too harsh. But as they say, you can't please all of the people all of the time.

Throughout, I have emphasized the creative side of evolution. There are two reasons for this emphasis. First, evolution and development are indeed the two great processes of biological creation. One has created organisms with trillions of cells from a single-cell starting point over a period of about a billion years. The other repeatedly achieves the same thing over a single lifetime. They are, in my view, the two most awesome processes in the whole of biology.

Second, one way of summarizing the deficiencies of the 'modern synthesis' is that it is a theory based almost entirely on destructive agencies. For an overarching theory of evolution this seems rather inappropriate, given all the creation of novelty that evolution entails. But this sweeping generalization needs a closer look. What I mean is that the core of the modern synthesis is natural selection. In one of its forms – selective mortality – it is clear enough that it is a destructive force. Both of two variant types of organism in a population (to take the simplest case) experience mortality; but one dies more rapidly, in proportionate terms, than the other. So the population evolves by selective destruction. However, in the other form of selection – a differential in birth rates rather than in death rates – the destructive nature of the process is less immediately obvious. Nevertheless, it is still there. The reason is that if one variant out-breeds another, the less fit variant will not just be reduced in frequency; it will, eventually, die out altogether, leaving the population composed only of the other form. This happens because populations can't increase in number indefinitely. Rather, they eventually hit a 'ceiling' set by their food supply or some other ecological factor, and a differential in birth rates then becomes just as much a form of selective destruction as a differential in death rates.

* * * *

So it becomes clear that the introduction of the variation upon which selection acts needs to be a major focus of attention. The above selective scenario, like most, assumes the prior existence of variation. And I mean just that: it is an *assumption*, not a focus of investigative attention. But the introduction of the variation is a crucial part of the creative process that we call evolution. It should not be relegated to the role of an assumed footnote. It should be up there in the forefront of our theory. In the 'inclusive synthesis' it is (or will be, depending on how far you think we've come) in that forefront position.

In the end, it all comes down to what we regard as most interesting, and this is a subjective judgement that each of us makes for ourselves. Some palaeontologists have spent their working lives studying that ultimately destructive process of mass extinction. Personally, I would have to agree that the destruction of the majority of the Earth's biota in a short space of geological time, whether by asteroid impact or less dramatic terrestrial causes, is a fascinating topic. But I'm even more fascinated by questions like how animals without heads can evolve into animals with heads; and how those heads can diversify, producing the small brains of snails and the big brains of their octopus cousins; or the small brains of mice and the big brains with which we humans can contemplate our evolutionary and developmental origins.

17 Social creatures

The originality of 'new' scientific ideas is often disputed. Whatever you come up with, someone will say that somebody somewhere has said it before. It is probably true that several people had the idea of natural selection before Darwin, but did not elaborate the concept, back it up with substantial evidence, or write about it eloquently enough to make anyone pay much attention. Equally, the Copernican revolution in our view of the solar system – from Earth-centred to Sun-centred – was re-enacting a similar shift in Chinese thinking that occurred centuries earlier. So, what of the 'new' ideas put forward in this book? Do I claim that there are any, and if so, am I right? In particular, do I claim that my 'biased embryos' approach is novel? I'll try to answer these questions shortly.

For the benefit of any readers who have jumped ahead to the final chapter from somewhere near the outset (the 'book as a thick mud sandwich made with deliciously fresh bread' syndrome), and for revision purposes for everyone else, here is a single-paragraph re-cap of the 'biased embryo' view of evolution.

Natural selection is not the 'main' orienting agent of evolution as Darwin claimed. Rather, it is one partner in an interacting duo that *is* the main determinant of the direction of evolutionary change, inasmuch as 'main' is a meaningful term in a multilevel process extending all the way from molecular changes in genes to mass extinctions. The other partner is developmental bias; that is, the tendency of the developmental system of any creature to produce variant trajectories in some directions more readily than others. There are many possible creatures that natural selection could cause to spread in the absence of developmental bias; but some of these never appear. So in a sense developmental bias chooses an actual subset of creatures from the set of all possible creatures, and is thus as much responsible for the Earth's biota looking the way it does as is natural selection. Not a conflict, as you can see, but rather a partnership. It's just that one partner has got the lion's share of the publicity up to now, thus giving us a 'lop-sided' view of the evolutionary process. It is this lop-sided view that I have been trying to correct herein.

Before embarking on our final journey, it will be fruitful to make a brief historical digression to 1986, 1866 and beyond. In 1986, Stephen Jay Gould and Niles Eldredge, the architects of the 'punctuated equilibrium' theory that sees evolution as prolonged stasis interrupted by fleeting moments of change, published a short paper[1] entitled 'Punctuated equilibrium at the third stage'. They were referring to three stages in the progress of new scientific theories that had been noted by von Baer (in 1866), who himself was referring to an earlier statement by Agassiz. Stage 1 is a reaction that the new theory is false; stage 2 is the view that it is against religion; and stage 3 is the assertion that it is perfectly correct but that in fact we all knew it (whatever it is) all along, and that it is thus not new at all.

The purpose of this digression is not to invoke 'stage 2' (though I do comment very briefly on religion below), but rather to draw attention to the important question raised by the 'stage 1' reaction, namely, 'is the theory true?' This gives a second dimension to our journey in the present chapter. We can think in terms of a two-by-two table of possibilities where the row headings are 'new' and 'not new'; and the column headings are 'correct' and 'incorrect'. Of course, theories may be partially new and partially correct, but let's take the simpler, all-or-nothing approach as our starting point, and ask into which of the four boxes the 'biased embryos' approach falls (Figure 31).

* * * *

We'll start with the decision of which row – that is, 'new' or 'not new'. It is clear that many influential figures throughout the history of evolutionary biology have downplayed, or even rejected outright, the idea that developmental bias, or, if you prefer, the structure of the available developmental variation, has an important role to play in the control of evolution's direction. Although there are also many biologists who

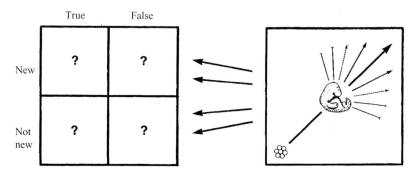

FIGURE 31 Four possible perceptions of the 'biased embryo' view of evolution: true/new; true/not new; false/new; false/not new.

have in one way or another promoted a role for developmental bias, they have not succeeded in persuading the community of biologists at large to perceive it as something of considerable importance.

Perhaps the biologists who have been most inclined to accept a role for development in evolution have been those who work explicitly at the organismic level. Included here are both comparative embryologists and palaeontologists. Many of these folk were excluded from the formulation of mainstream evolutionary theory during what I sometimes think of (with conceptual tongue in cheek) as the 'dark ages' between about 1900 and 1980. Of course, there were exceptions, as I have already mentioned, such as Gavin de Beer and G. G. Simpson, but few would now deny that these were playing second fiddle to the population geneticists, for whom the population and the gene, rather than the organism and its development, were the prime focus of attention. As I have said before, it is not my aim to try to dislodge population genetics from its acknowledged position as a major contributor to evolutionary theory; rather, I wish to acknowledge its importance, but to persuade it in turn that it has an equal partner without which we will never reach the *complete* theory of evolution that we all seek. Given the healthy heterogeneity within most academic disciplines that I have referred to earlier, some population geneticists may require little if any persuasion, others rather more.

Let's look at a palaeontological contribution to the debate about evolutionary mechanisms that has appeared since the end of the 'dark ages' (in 1997 to be precise) and has been aimed at a general audience.[2] I am referring here to the book entitled *Shapes of Time*, written by the British palaeontologist Kenneth McNamara.

McNamara uses a wonderful culinary analogy to try to put the organism back in a centre-stage position in evolutionary theory. He pictures the genome as a chef, the organism as a meal, and natural selection as the customers in a restaurant. His view of the omission of the organism from the core of evolutionary theory for much of the twentieth century is as follows: 'It's as if a food critic were concerned simply with the characteristics of the chef and the peculiarities of the customers, and took virtually no notice of the food and how it was produced.' You will recognize this as corresponding to the central point of one of my earlier chapters ('The return of the organism'). However, in my view McNamara does not go quite far enough, because he concentrates almost exclusively on heterochrony. Although this is important, it is only one out of four possible types of developmental reprogramming, as I noted in Chapter 7.

It is not just palaeontologists who accept an important evolutionary role for development. The other group that I need to refer to in this context is the subschool of population genetics *sensu lato* that goes by the name of quantitative, or biometrical, genetics. The difference between this and population genetics *sensu stricto* is that the former is concerned with the evolution of complex phenotypic characteristics to whose variation many genes contribute, while the latter has often focused, especially in its early stages, on single-gene polymorphisms.

During the period from about 1930 to 1970, when the mathematical framework of evolutionary theory was being put together, studies of polymorphism in populations, both theoretical and empirical, were more influential than the sorts of study typically conducted by quantitative geneticists. I don't claim to know exactly why this was the case, though it probably has something to do with the difficulties inherent in studying partially heritable characteristics in wild populations. As a representative of that cohort who were research students in the 1970s, I 'grew up' thinking that quantitative genetics was more related to animal and plant breeding than to evolution. This, I now realize, was a seriously misguided view; in fact, quantitative genetics is highly relevant to both.

There is, however, a problem. Quantitative genetics is a difficult subject. It requires a degree of mathematical ability that most biologists (and most laypeople) lack. And its practitioners have a habit of taking no prisoners. Almost all of the books and papers that I have read in this area have been hard going. Let me give you an example. The geneticist James Cheverud,[3] writing in *Journal of Theoretical Biology*, says that 'the genetic variance/covariance matrix of quantitative genetic theory measures developmental constraints'. Most people's reaction to this will be something along the lines of 'what on earth is a variance/covariance matrix?' In fact, it is a way of measuring how different characteristics of organisms tend to co-vary. That is, it is a way of representing developmental bias – not just constraint. So there is an important message lurking in statements such as Cheverud's, but it tends to reach only a select audience because of the technicality of the language.

So in one way, my 'biased embryos' approach is not new in that it has been adopted, albeit in a rather mathematical manner that is inaccessible to many biologists, by quantitative geneticists. In another way it might yet be new, because I am interested not just in being able to quantify the *results* of developmental bias, for example at the adult stage. I am interested also in the *reasons* why some variant developmental trajectories are more probable than others. Clearly, most quantitative geneticists regard this as a topic that lies within someone else's field of enquiry. So who are these other people?

It should come as no surprise that they are students of evo-devo. And they are students of two distinct types. First, there are those who you might describe as theoreticians, who have focused their attention on developmental bias and related issues. Second, there is the new generation of molecular and cellular 'evo-devologists', who seek explanations for comparative developmental phenomena at the level of genes, their products, and other interacting subcellular entities. The gradual convergence of these theoretical and empirical sides of evo-devo is a major goal for the future.

So, the nature of my 'biased embryos' approach is finally becoming clear. To the question 'is it new?' I can, appropriately, give the famous Irish answer 'yes and no'. It is not new because all the ingredients are 'out there', lurking in the literature, some of them well-known, others specialist and obscure. But my attempt at a *synthesis* of these various ingredients is, like other such attempts by other authors, idiosyncratic, and thus at least 'new in parts' (with apologies to the proverbial curate's egg).

*　*　*　*

But there is a risk that the 'new' theory advanced here may be wrong; and that those who have claimed that natural selection alone determines the direction that evolution takes, of whom Alfred Russel Wallace may well have been the first, will turn out to be right after all. It is to this issue of validity rather than novelty, or columns rather than rows in my earlier analogy, that we now turn.

It makes sense to approach this issue with three distinct questions in a logical order. The first is: does developmental bias exist in the real world as well as in my (and others') imagination? I am confident that the answer to this question is an unqualified 'yes'. Such an answer is possible on the basis of a single clear example. And one such example, though a rather unusual one, is the fact that all known species of centipedes have odd numbers of leg-pairs (between 15 and 191). Not a single species has an even number. If there were only three species of centipede, we could write this off as being due to chance; but since there are in fact about 3000 species, this explanation is clearly untenable.

The second question is: is developmental bias the norm or the exception in nature? I am confident that the answer is an unqualified 'norm'. It is not possible to reach this conclusion on the basis of a single example, because the question is now about relative frequency. So I will take a different, and more abstract, kind of approach. Consider any particular type of creature whose developmental trajectory can change in a certain number of ways – let's say 100. Does the probability of generating variant trajectories (ultimately via mutation and reprogramming) vary across these 100 directions, and if so, *how* does it vary?

Regrettably, we can't specify a comprehensive pattern of variation yet for any real creature, but despite that I think we can say that a variable pattern of some sort is vastly commoner than the alternative – equiprobable change in all 100 directions. The reason is simply that equiprobability is only one of an almost infinite series of patterns. All the others represent, by definition, various types and degrees of developmental bias. Therefore, bias would, on an *a priori* basis, be expected to be the norm rather than the exception in nature.

Now we come to the third and hardest question: does developmental bias play an important role in determining the direction of evolutionary change? The fact that it exists, and is almost certainly widespread, does not in itself allow us to give a clear answer to this final question.

If a population is evolving on the basis of its standing variation, then, while absolute bias *necessarily* contributes to the array of creatures that evolution produces, relative bias only *might* do so. That's a more complex statement and we need to dissect it. The problem is this. In the examples that I gave in Chapter 8, I deliberately used adaptive landscapes that showed how bias could have a major effect. Given my aim in this book, it would have been foolish to do otherwise. But now we need to take a step back and consider the question of what adaptive landscapes are really like in nature.

The difficulty that we encounter here is that we simply don't know what they are like. Various authors have postulated gentle landscapes, rugged landscapes, even landscapes with 'holes' in them. But these are all just mental game-playing. The only things that we know for sure about real adaptive landscapes are: they are many and varied; and they change over time for all sorts of reasons, the simplest of these being changes in the prevailing environmental conditions. But as to the prevalence of particular 'shape classes' (e.g. gentle versus rugged) we know little. That is important because some types of adaptive landscape, when combined with some types of variation, may render developmental bias irrelevant to the direction of evolutionary change.

So now the nature of my 'biased embryos' approach becomes even clearer. It is a *hypothesis* rather than something that we know to be true. Given some types of adaptive landscape (or 'seascape'), the interacting duo of bias and selection will indeed set the evolutionary sails; but in other types of landscape, selection may do the job alone, despite the existence of widespread bias. It all depends on the relative commonness of different types of landscape, different types of bias, and the mapping between the two. This must surely be a major target for future study.

Now to that other kind of evolutionary change, where a population is 'waiting for mutation' – so we're back to that old Goldschmidtian 'can of worms' that we encountered in Chapter 9. Perhaps the main reason why few biologists have agreed with Goldschmidt's view that evolution proceeds through large macromutational leaps is that he proposed his theory in such an extreme form.[4] And perhaps his 'extremeness' can be held at least partly responsible for the retreat towards the other extreme that followed, in which many biologists sought to explain all evolutionary changes in terms of tiny contributions from many (perhaps hundreds) of genes. It is only in the last couple of decades that we have come to realize that the truth lies somewhere in between. Analyses of the genetic bases of characters like body size and shape that vary in a continuous way have revealed that in many cases relatively few genes are involved, and that often one or two of these have much more major effects on the character concerned than the others. This takes us into the realm of 'meso-mutations'.

A population that has a small amount of standing variation that is insufficient to climb even the closest fitness peak is truly waiting for mutation (or immigration sometimes). If a mesomutation comes along, it may well determine which one of a series of previously unreachable peaks comes within the domain of reachability. In this kind of evolutionary process, the role of developmental bias is less sensitive to the precise nature of the adaptive landscape, and the general expectation would be that the 'interacting duo' would determine the direction in which evolution proceeds.

*　*　*　*

It's about time I got around to explaining the reason for my up-to-now cryptic choice of title for this, the book's final chapter. Science is an intellectual endeavour. Indeed it is a hugely exciting intellectual endeavour that has, at its heart, the formulation of general theories about life (biological sciences) or other recognizable domains, from subatomic particles to the universe itself (physical sciences). It is a sad fact that this excitement can sometimes be hidden behind uninspired teaching or turgid writing. But, such screens aside, the excitement is very real, and I find it hard to imagine anything that is quite equivalent. We sit here in a largely alien cosmos, with a long and mostly unknown history, having started from a beginning that, if current 'big bang' views are correct, seems bizarre in the extreme. Science is a way of trying to understand who we are and where (and when) we have come from. It is a unique way of trying to reach such an understanding, characterized by the honesty that is implicit in a methodology that has the inbuilt feature of rejecting previously accepted 'truths' if new evidence makes it clear that we should do so.

In this respect, science is the opposite of 'faith', which involves clinging to an old set of beliefs, however strong the evidence against them might become. This is not to say that science should supplant religion; personally, I don't think that it should. But it does mean that all faith-based religions should temper their false certainty with a little of the scientist's agnostic honesty. There is a *dis*honesty inherent in 'fundamentalist' stances that all too easily lends itself to intolerance of the views of others, sometimes with appalling consequences.

Science is a social venture as well as an intellectual one. It is done in society as well as in the mind. Interactions between people have a crucial role to play in deciding which theories become accepted (or even generally known about) and which do not. The fact that Gregor Mendel's discovery of the particulate nature of inheritance remained unknown to most scientists for more than thirty years had little to do with the quality of Mendel's work. But it had a lot to do with where he worked, the people he was (and was not) in regular contact with, how he chose to entitle his posthumously famous paper,[5] and where he decided to publish it. We humans are interacting, social creatures, but there are limits to the number of interactions that each of us is capable of participating in. We are necessarily selective in whom we talk to and what we read. Life is too short for any other type of approach; and even if it wasn't, I'm not sure that a totally unselective approach would make any sense.

What all this means is that those of us who are engaged in scientific research, and in the spreading of its findings to the widest possible audience, have an obligation to consider the social dimension of science, and to avoid wasting time on writing things that will rarely if ever be read. We need to be astute in how we choose to 'broadcast' our cherished theories. Of course, if the theories are wrong this will speed their demise just as surely as it will speed their acceptance if they are right. But we should rejoice in both effects, for both represent progress in our understanding of life, the universe, and everything. As well as collectively revelling in that particular form of honesty that the scientific approach embodies, we must each personally have the humility to put this into practice if it becomes necessary in relation to whichever theory we hold dear. So I cast my biased embryos to the scientific and social world at large, and await its verdict with interest.

Glossary

All definitions are given as concisely as possible. To look up a two-word phrase like 'developmental bias' or 'natural selection', use the beginning of the phrase, even if it is an adjective. Words within any of the definitions that are given in *italics* are themselves defined in the Glossary.

adaptive landscape A pictorial way of showing how *fitness* varies with different values of one or more *characters* (e.g. body size).

amino acid A building block of a *protein*. There are 20 or so different kinds of amino acid. Different proteins (e.g. *haemoglobin* or a digestive *enzyme*) differ both in their total number of amino acids and in their amino acid sequence. A typical protein *molecule* has a few hundred amino acids.

amphioxus A primitive *chordate* that has the appearance of a rudimentary fish lacking well-developed head, paired fins and tail. The vertebrates are thought to have arisen from an ancestor of this general sort.

annelid A segmented worm. Annelida is a *phylum* of animals with many thousands of species, one of the most familiar being the earthworm.

arachnid A group of animals that includes the spiders and scorpions. The arachnids are a subgroup of *arthropods*.

Arthropoda The largest animal *phylum*, whose members are characterized by a hard exoskeleton and jointed legs. There are four main arthropod subphyla or classes: the insects; the *crustaceans*; the *myriapods*; and the chelicerates (i.e. the *arachnids* and their close relatives). There are also some extinct arthropods that do not belong to any of these groups, of which the best known are the *trilobites*.

artificial selection A term equivalent to *natural selection* when the selection is carried out by a biologist or an animal/plant breeder rather than by natural agencies (e.g. a predator).

Bilateria A high-level *taxon* that includes the bulk of the animal kingdom. As the name suggests, the bilaterian *body plan* has a head end, a tail end, a left side and a right side. That is, it is bilaterally symmetrical. Those primitive animals whose *lineages* split off from the main animal stock before this design arose lack bilateral symmetry, for example the sponges and jellyfishes.

billion One thousand million. (The old 'British' billion, which was a million million, is not used in scientific work.) The Earth is about 4.5 billion years old.

biosphere The thin envelope around the Earth in which most *organisms* are found. It extends from under the soil (or sea floor) to a few hundred metres up in the atmosphere.

blastula An early embryonic stage found in some kinds of animals (including many vertebrates). It consists of a small number of *cells* organized into an approximately spherical shape, with an internal cavity of variable size.

body axes The three main axes of a *bilaterian* animal, namely anteroposterior, dorsoventral and left–right. Alternatively, whatever other axes can be substituted for these in an *organism* that is not bilaterally symmetrical.

body plan The general layout that characterizes body structure in a high-level group of *organisms*, such as a *phylum* or subphylum. For example, our own (vertebrate) subphylum is characterized by an internal skeleton.

Cambrian The period of geological time from about 540 million years ago (*MYA*) until about 490 MYA. This was the first period within the Phanerozoic Era (540 MYA to the present). There are relatively few animal fossils before the Cambrian, which, in contrast, is associated with an 'explosion' of fossils.

cell The basic building block of an *organism*. The simplest organisms (e.g. bacteria) consist of just a single cell. Large animals and plants, in contrast, can possess many *trillions* of cells. Each cell consists of a *nucleus* surrounded by *cytoplasm* that is in turn surrounded by the cell membrane.

character Any recognizable feature of an *organism* that is potentially capable of varying among a series of different character states. Characters can be described at a variety of levels from the *molecular* (e.g. *DNA* sequences) to the *morphological* (e.g. segment number or body size).

chordate An animal with a structure called a notochord, which is a stiff skeletal rod running along the *dorsal* midline either for a transient period during development of the *embryo* or for a more protracted period, up to the complete lifespan. The vertebrates are the biggest subgroup of chordates, but there are also some more primitive chordates such as *amphioxus*.

chromosome A thread-like structure, visible during cell division using an ordinary microscope, that contains many (usually thousands of) *genes*. Each species has a characteristic number of chromosomes (e.g. 23 pairs in humans). Chromosomes are found inside the *nucleus* of each *cell*.

clade All the descendants of a particular (usually unknown) ancestral species. The vertebrates constitute a clade; so do the mammals and the birds; but the reptiles do not, because the group 'Reptilia' excludes some of the descendants (namely mammals and birds) of the original, ancestral reptile.

clone A *population* in which all the *organisms* are genetically identical. This can arise as a result of asexual reproduction. (Note that 'clone' is also used in a different way, as both a noun and a verb, in many of the techniques of molecular biology.)

coadaptation A term describing pairs or groups of *characters* that have evolved so that they function in an integrated manner. A whole *organism* can be thought of as a highly complex coadapted entity consisting of many characters.

complexity The number of components, or types of components, that an entity such as an *organism*, a *cell*, or an automaton is composed of. So, for example, a vertebrate, with perhaps 200 cell types, is a more complex organism than a jellyfish, which has far fewer cell types.

congeneric A term applied to species that are sufficiently close relatives that they belong to the same genus – for example the 'ordinary' chimp and the pygmy chimp; or the blue tit and the great tit.

convergence The evolutionary process in which two (or more) *lineages* that started off different from each other end up becoming more similar, at least in some *characters*. For example, the external body shape of mammals that returned to water (e.g. dolphins) has converged with that of fish.

contingency See *historical contingency*.

co-option The evolutionary process in which something (often a *gene* or a group of interacting genes) that starts off with a particular role ends up acquiring a new role either as well as or instead of its original one.

co-variation The situation in which two *characters* tend to vary together either between *organisms* within a species or between different species. For example, humans with longer arms also tend, on average, to have longer legs. Equally, a species of mammal that has longer forelegs than another usually has longer rear legs too.

Crustacea The large group (sub*phylum*) of *arthropods* that includes crabs, lobsters, shrimps, prawns, water-lice and woodlice.

cryptic Hidden/hard to see. Often used to refer to the colour patterns or shapes of animals (e.g. a moth's 'camouflaged' pigmentation pattern or a stick-insect's 'twiggy' body form).

cytoplasm The semi-fluid material within a *cell* that fills the space between the inner *nucleus* and the outer cell membrane.

Darwinism The school of thought that is based on Darwin's own view that *natural selection* is the main, but not the only, agent of evolutionary change.

developmental bias Non-randomness in the variation (both discrete and continuous) in *developmental trajectories* upon which *natural selection* acts. Includes both *developmental constraint* and *developmental drive*.

developmental cascade The sequence of interactions among many *proteins* and other *molecules* that governs the way in which a *developmental trajectory* proceeds.

developmental constraint The difficulty or impossibility of producing certain developmental variations from a given starting point. These are often referred to respectively as relative/quantitative constraint and absolute/qualitative constraint.

developmental drive The ease of producing variation in development in particular directions. Thus opposite, and complementary, to *developmental constraint*.

developmental reprogramming The deflection of a *developmental trajectory* over all or any part of its course caused by mutation in one or more of the genes that help to control the direction that the trajectory takes. Unlike *developmental bias*, a 'theory-neutral' term that is simply a description of what happens.

developmental trajectory The series of forms that an organism passes through as it progresses from its starting point (usually a fertilized egg) to its adult state.

direct development The type of *developmental trajectory* in which the *organism* approaches its adult condition gradually (as in humans) rather than via some *larval* form that is markedly different from the adult (e.g. the tadpole stage in frog development).

disparity The amount of difference between two *organismic* forms. Normally used in cross-*taxon* comparisons. For example, the degree of disparity between a human and a dog is greater than that between a human and a chimp.

DNA Deoxyribonucleic acid. The *molecule* of which *genes* are made. Its structure is that of a 'double helix', as famously revealed by Watson and Crick in 1953.

dorsal The 'back' of a bilaterally symmetrical animal. Often used in the form dorsal–ventral (or dorsoventral), to denote one of the three *body axes* (the others being anteroposterior and left–right).

downstream A term applied to the position of one component in a *developmental cascade* relative to another. So if component X activates component Y, which in turn activates component Z, then Y and Z are both downstream of X; and Z is also downstream of Y.

embryo An early part of most *developmental trajectories* in which the developing *organism* is protected from the outside environment by virtue of being enclosed in (for example) a mammalian womb or a bird's egg.

enzyme A type of *protein* that facilitates one of the body's many chemical reactions. Usually the product of a single *gene*; but some enzymes are more complex and are the products of two or more genes.

epigenesis A term sometimes used as a synonym of development. But also used in the sense of the non-genetic aspect of development. So we can think of the *developmental trajectory* as something that results from the interplay between a genetic and an epigenetic programme. Previously used as an opposite to the now-discredited theory of 'preformation' wherein a sperm (or egg) cell was thought to contain a tiny version of the adult.

epiphyte A plant that grows on the surface of another *organism*, whether plant or animal. For example, the algae that grow on pondweed or on the shells of pondsnails.

evo-devo Evolutionary developmental biology. The new (*c.*1980–) 'hybrid' discipline whose proponents study the interrelationship of the two great processes of biological creation – evolution and development. A modern-day equivalent of nineteenth-century comparative embryology.

fauna All the animals in a given geographical area. Used, for example, to describe the range of animals found in a particular country (e.g. the British fauna) or a particular region or type of region (e.g. alpine fauna).

fitness The combined survival and breeding probabilities of one variant *organism* in a *population* compared with another. The popular phrase 'survival of the fittest' equates with natural selection.

fixation The process in which a *gene* that is spreading through a *population*, often under the influence of *natural selection*, finally reaches 100 per cent; thus sending the other version of the gene into oblivion.

flora The plant equivalent of *fauna*. That is, all the plants of a particular area; as in 'the British flora' or 'alpine flora'.

fossil record The whole range of fossils across all *taxa* and over all periods of geological time; or, if qualified appropriately, some segment of this.

ganglion A smallish concentration of nerve *cells* in a particular part of the body. In some simple animals, a ganglion at the head end constitutes a sort of primitive brain. In more complex animals, the brain is too large to be called a ganglion, but there may be ganglia elsewhere in the body (e.g. in every segment, in many segmented animals).

gastrula A developmental stage that most animals go through, in between the *blastula* (or equivalent), and later stages in which the individual organs begin to be formed. Usually accompanied by much *cell* migration. Establishes the main *body axes*.

gene A functional unit of *DNA* that usually makes a *protein*. In a few cases, genes make partial proteins or what might be thought of as protein precursors that never become proteins (RNA *molecules*). A typical animal or plant gene will extend over many thousands of *nitrogenous bases*.

gene duplication Any process occurring in the *genome* where one or more *genes* end up being present as double copies. Includes localized 'tandem' duplication where the duplicate copies end up sitting beside each other at a particular spot on a particular *chromosome*. Included in broad-sense, but not narrow-sense, *mutation*.

gene expression The functioning of a *gene*. That is, its making of its product in those places, and at those times, where/when it is switched on. Often used in the phrase 'gene expression pattern', which relates to the regions of developmental space and time for which the gene concerned is active.

gene frequency The frequency of a particular version of a *gene* relative to its alternative(s) in a *population*. Can be altered by *natural selection* or *genetic drift*.

genetic assimilation The process by which a *phenotypic character* that initially only appeared in *organisms* subjected to some environmental stimulus eventually appears independently of that stimulus. Caused by *natural selection* or *artificial selection* for *genes* that increase the probability of the developmental system producing the character concerned.

genetic code The system of building a *protein* by using sequences of three consecutive *nitrogenous bases* within a *gene* to code for particular *amino acids* in the protein. Which triplet of bases codes for which amino acid is almost universally set throughout the living world. In many cases, two or more triplet codes produce the same amino acid. This is referred to as *redundancy*.

genetic drift The random fluctuation of *gene frequency* due to chance events, in contrast to the systematic change in gene frequency that can be caused by *natural selection*.

genome The totality of genetic material in any particular *organism*, often described in terms of the approximate number of *genes*. For example, the recent 'human genome project' has revealed that our own genome has approximately 35 000 genes – about half of what many people had thought.

genotype The combination of two copies of any particular *gene* that an individual animal or plant has by virtue of its inheriting one copy from its mother and one from its father. (If the two copies are the same, the genotype is said to be homozygous; if different, heterozygous.)

haemoglobin The *protein* found in the red blood *cells* of mammals (and some other animals) that carries oxygen from the lungs to the various tissues of the body.

haplodiploidy A breeding system in which one sex, usually the female, carries two copies of each *gene* (diploid) while the other, usually the male, carries a single copy (haploid).

heterochrony Literally 'other timing'. Refers to situations in which the rates of development of two or more parts of an *organism* (e.g. organs, limbs) are evolutionarily shifted relative to each other.

heterometry Literally 'other amount'. An evolutionary change in the amount of some developmental entity (such as the product of a developmental *gene*); as opposed to a change in its type, location, or timing.

heterotopy Literally 'other place'. The spatial equivalent of (temporal) *heterochrony*. Refers to situations in which some developmental event is evolutionarily shifted from one part of the developing *organism* to another.

heterotypy Literally 'other type'. An evolutionary change in development that is more than just a change in the amount, location or timing of something that is already there. Rather, something 'novel' has been produced.

historical contingency The role of 'one-off' chance events that have major effects on the course of evolution. For example, the non-production of a whole *phylum* of animals due to the extinction, as a result of some unpredictable geological event, of what would have been the 'stem species' of the phylum concerned. (Cf. *stochastic* processes such as *genetic drift* – repeated minor chance events.)

homeobox A *DNA* sequence of approximately 180 *nitrogenous bases* that is found in many developmental *genes* in animals and plants. The corresponding bit of the *protein* produced by such a gene binds to other *downstream* genes to switch them on or off, or to regulate the rate of their expression.

homology A term invented in pre-'Origin of Species' times, but now given an evolutionary interpretation. When the partial similarity of a component, whether a *gene* or an organ (etc.), of one type of organism with its counterpart in another is due to shared ancestry rather than convergence, the components are said to be homologous.

hopeful monster A phrase introduced by Richard Goldschmidt to describe a 'monster', that, regardless of its gross differences from its parents (due to mutation), manages to establish itself and becomes the basis for the sudden origin of a new higher *taxon*. Now generally thought of as a rather improbable form of evolutionary change.

hormone A type of *molecule* that is produced in a particular place in a multicellular organism, but that circulates widely in the body and is thus capable of producing widespread developmental effects.

housekeeping gene A *gene* whose product is involved in the general 'housekeeping' of the *cell*, for example in those cellular activities that provide the energy that the cell (and the *organism*) needs in order to survive. This would be a distinct category from that of 'developmental genes' were it not for the fact that many genes have dual/multiple functions.

Hox gene A type of *homeobox*-containing *gene* that governs patterning along the anteroposterior *body axis* of a bilaterally symmetrical animal. Hox genes are often found in clusters on particular *chromosomes*.

imaginal disc An approximately disc-shaped piece of tissue found in the *larvae* of certain groups of insects (including flies) that will form part of the adult during *metamorphosis*. Discs are usually found in pairs, and different discs occur in different segments. (For example, the pair of fly wing discs is located in the second thoracic segment of the larva.)

indirect development A *developmental trajectory* in which the adult stage is reached indirectly via a *larval* form that is distinctly different from what follows it (e.g. a tadpole or caterpillar).

internal selection A phrase brought into usage by L. L. Whyte in the 1960s to describe a particular type of *natural selection* in which the differences in *fitness* among *organisms* in a *population* arise as a result of varying degrees of internal integration rather than varying degrees of environmental adaptation (external selection). These two types of selection are better thought of as the opposite ends of a continuum of possibilities rather than as discrete alternatives.

Lamarckian Used to describe evolutionary mechanisms involving the inheritance of acquired characteristics, which result in evolutionary change being determined by *use and disuse*. Now generally discredited.

larva A developmental stage that occurs in the *life cycles* of some animals (e.g. frogs, butterflies, many marine invertebrates) between hatching and adulthood. It is characterized by being entirely different in appearance from the adult.

life cycle The whole series of developmental and adult forms in between one stage (e.g. the fertilized egg) and the same stage one generation later.

lineage An evolutionary line of descent. Usually thought of as starting at some particular branch point (or divergence) in an evolutionary tree, and leading to some particular creature.

macromutation A *mutation* with a large effect on the development of an *organism*. As there is no clear separation of 'large' from medium or small, it is better thought of as one end of a spectrum of possibilities.

Mazon Creek The name given to an area in Illinois and to the abundant and often well-preserved fossils of Carboniferous age (*c.* 300 *MYA*) that are found there.

mesomutation A term used herein to refer to *mutations* that are intermediate in their magnitude of effect on the development of an *organism* between *micromutations* and *macromutations*.

metamorphosis The developmental process through which a *larva* turns into a juvenile or an adult. Sometimes this process is protected within a hard outer casing (for example the chrysalis of a butterfly).

micromutation A *mutation* that has a small effect on the development of an *organism*. Like its opposite, *macromutation*, it is better thought of as one end of a spectrum of various magnitudes of effect, rather than as a distinct category.

modern synthesis The synthesis of evolutionary theory that was achieved in the mid twentieth century, as a result of the integration of findings from several disciplines, notably genetics, ecology, *systematics* and *palaeontology*. But largely lacking a developmental component, and so only a partial synthesis.

module Any quasi-autonomous part of an overall developmental process. For example, a limb bud or a segment.

molecular drive An umbrella term for the directional evolutionary effects of several different molecular processes occurring within the *genome*.

molecule The basic unit of a chemical compound. Where these basic units are very large (for example in the case of *DNA* or *proteins*) they are often referred to as macromolecules.

mollusc Any member of the animal *phylum* Mollusca. Examples include snails, slugs, bivalves and octopuses.

morphology The study of *organismic* structure or form. Often also used in a loose sense to refer to organismic form itself, sometimes in a broad way, but sometimes in a narrower way, with a focus on external form (in which case internal structure or its study is referred to as anatomy).

mutation A change in the sequence of *nitrogenous bases* within a *DNA molecule*. In its simplest form, a change in just one base in one particular *gene*, but many more complex forms of mutation are also known.

mutation bias A situation in which *mutations* occur more readily in some directions than others. In certain circumstances, this can affect the direction of evolutionary change.

MYA Millions of years ago. Sometimes abbreviated instead to MYBP (millions of years before the present).

myriapod A group of terrestrial *arthropods* characterized by possession of many (ten or more) pairs of legs. The most familiar examples are centipedes and millipedes.

natural selection The process in which those genetic variants in a *population* that have a higher probability of survival and/or a higher reproductive rate tend to spread through the *population* at the expense of other variants.

neo-Darwinism The school of thought that began to emerge in the 1930s with the advent of mathematical models of *natural selection*, and that is associated with the development of the *modern synthesis* of the 1940s, 1950s and 1960s. Neo-Darwinists are those who consider *natural selection* to be the main or sole determinant of evolutionary direction.

nitrogenous bases Those building blocks of a *DNA molecule* whose sequence determines, through the *genetic code*, the type of *protein* that a *gene* will produce.

nucleus The central membrane-bound body found within an animal or plant *cell*. The nucleus contains the genetic material, in the form of a number of pairs of *chromosomes*, each typically carrying thousands of *genes*.

ontogeny The *developmental trajectory* of an *organism*, including embryonic and post-embryonic stages.

organism An individual creature. For example, a human, an oak tree, a snail or a lizard. Hard to apply in some cases, for example when organisms live in colonies or reproduce vegetatively.

orthogenesis An evolutionary trend in a particular direction (e.g. larger body size), thought to be determined internally rather than externally. Supporters of this view have ranged from mystics to those who seek internal mechanisms for such trends.

palaeontology The study of fossil *organisms* of all kinds. Usually this involves fossilized body parts (skulls, shells, legs etc.). But it also includes the study of fossilized remains of animal activities, such as worm-holes and footprints (known as trace fossils).

pan-externalist A type of *pan-selectionist* who believes that *natural selection* in response to external, ecological factors is the overwhelming determinant of the direction of evolutionary change.

pan-selectionist Someone who believes that the overwhelming or even sole determinant of evolutionary direction is *natural selection*. The extreme wing of *neo-Darwinism*.

phenotype The type of *organism* produced as a result of possession of particular versions of one or more *genes*. Includes all *characters* except the genes themselves. Studies of phenotypes often focus on particular characters (e.g. body size, shape, pigmentation patterns).

phylogeny A pattern of evolutionary relationship among three or more *taxa*. An evolutionary tree may be referred to as a phylogenetic tree, and those biologists who focus on attempting to discern the true tree of relationship for a particular group of creatures are practising the discipline of phylogenetics.

phylum One of the great groups of animals, such as the *arthropods*, *chordates* or *molluscs*. The animal kingdom is thought to consist of about thirty-five phyla (plural). However, like most other levels of *taxon*, there is no universally agreed set of guidelines as to what determines that a group of animals should be called a phylum rather than (for example) a class.

placenta The mother's tissue that nourishes the embryo as it develops in the largest group of mammals (which are thus called the placental mammals). Absent in the most primitive mammals, such as the platypus or the kangaroo.

plankton The great mass of tiny aquatic creatures that floats below the ocean surface. Includes both plants (phytoplankton) and animals (zooplankton). Plankton includes the adult stages of some creatures and the developmental (*larval*) stages of others, for example *trochophore* larvae.

plasticity The production of changes in *organismic* form by direct environmental influences. For example the production of smaller adult insects when less food is available at the *larval* stage; or the production of different kinds of leaf above and below water in an aquatic plant.

pluteus A small marine *larval* stage that is part of the life cycle of many echinoderms (the group containing starfish, sea urchins and their allies). Similar in size but not in morphology to the molluscan *trochophore* larva.

population All the *organisms* of a particular species living in a particular area. For example, the human population of Iceland; the red deer population of a Scottish island; or the edible snail population of an area of French countryside.

pre-Cambrian A term for all periods of geological time before the *Cambrian* period that started approximately 540 *MYA*. The pre-Cambrian is characterized by a much sparser *fossil record* than the time since 540 MYA (which is collectively known as the Phanerozoic).

primordium The rudimentary early stage of a structure or organ that will become much more fully elaborated as development proceeds. For example many early animal *embryos* have limb primordia.

protein The type of *molecule* that is made by *genes*. There are many different types of protein. One of the largest categories is the *enzymes*. Some types of protein have major roles in development – for example those that switch genes on or off, or regulate their level of expression.

recapitulation The progressing through developmental stages or features resembling ancestral *organisms* by the *embryos* of descendant ones. For example, the transient appearance of gill clefts in human embryos. Always the resemblance is incomplete; and normally it is a resemblance to ancestral developmental, not adult, stages.

redundancy The duplication (or replication) of developmental control systems so that development will function normally even if one system fails. (The same approach is used in aircraft design.) Redundancy is also used in a different way to describe the *genetic code*, since more than one sequence of three *nitrogenous bases* will often produce the same *amino acid*.

reproductive isolation The separation of two or more genetically different *populations* of a single species (hence able to interbreed) into different *species* in their own right (hence unable to interbreed). Isolation can occur for many reasons, including differences in courtship behaviour or in the structure of genitalia.

segment polarity group The group of *genes* that are expressed in segmental bands in fruitflies and other *arthropods* and that help to determine which part of each segment is anterior and which is posterior.

selfish gene An expression brought to prominence as the title of a book by English biologist Richard Dawkins. There are various aspects to the selfish gene concept, one of the most important of which is the emphasis on *natural selection* spreading *genes* if they help their bearers (or in some cases close relatives). Whether this 'selfish' process ends up benefiting the species as a whole is an open question.

species selection The equivalent of *natural selection*, when the entities that are being selected are species rather than individual *organisms* (or *genes*). *Reproductive isolation* (and so the origin of new species) substitutes for birth; and extinction substitutes for death. So if two species have equal extinction probabilities but one has a higher probability of splitting, then that one will leave more descendant species.

stochastic A term referring to small fluctuations caused by random processes and having particularly large effects in small *populations*. For example, the human sex ratio is close to 1:1, but one family can easily have 100% of the offspring being of one sex, even if there are four or more children. See also *genetic drift*. (Cf. *historical contingency*: one-off events with major effects rather than an ongoing series of small changes.)

substrate The substance on or in which an *organism* lives. For example, soil, sand, rock. In the case of leaf-mining insects, the substrate is plant tissue; and in the case of animal parasites it is animal tissue.

synapsid A type of *tetrapod* vertebrate characterized by a particular skull structure. Includes the 'mammal-like reptiles'.

systematics The discipline concerned with the structure of the living world, in terms of the pattern of relationship between different types of creatures. Systematics pre-dated evolutionary theory; but nowadays evolution provides the context for systematic work.

systemic mutation A type of *macromutation* that was proposed by geneticist Richard Goldschmidt in the 1940s as a way of forming new types of animal 'all at once'. Now generally discredited.

taxon A grouping of related creatures. There are various levels (or ranks) of taxon (plural taxa) from the species up to the *phylum* and kingdom.

tetrapod A land vertebrate. Literally means 'four legs'. But includes humans, birds, and even snakes. Essentially the tetrapods are the result of the radiation of vertebrates that invaded the land from a quasi-amphibian ancestor, whose own ancestor, in turn, was a type of fish.

trait A term virtually synonymous with *character*, but probably originated from the study of behavioural rather than *morphological* characters.

trillion One thousand *billion*. Or 10 to the twelfth power, for those who like orders of magnitude given as powers of 10. The human body is thought to contain about 100 trillion *cells*.

trilobite A type of extinct *arthropod*. These marine creatures left a substantial *fossil record*. They probably represent a distinct class of arthropods, none of whose members survives today.

trochophore A type of early *larval* stage found in many marine species of *molluscs* and *annelids*. Small, semi-transparent and with bands of little hairs called cilia. Found among the *plankton*.

upstream A term applied to the position of one of the interacting components (e.g. a *protein*) relative to others in a *developmental cascade*. An upstream component can have an effect on a *downstream* one but not vice versa. However, in practice, some developmental *genes* get switched off and then on again. So the product of one of these can have both upstream and downstream roles.

use and disuse The phrase that Darwin used to refer to *Lamarckian* evolutionary processes. These involve a *character* becoming elaborated in the course of evolution by its frequent use, coupled with the inheritance of acquired size increases, shape changes (etc.); or conversely a character becoming 'vestigial', like the human appendix, due to lack of use. However, the lack of evidence for such processes has led to the abandonment of this idea by evolutionary biologists.

ventral The underside of a *bilaterian* animal (or the front in those, like ourselves, that walk upright). Defines one end of the dorsoventral *body axis*.

References

References to Chapter 1

1. Bonner, J. T. (1974). *On Development.* Cambridge, MA: Harvard University Press.
2. Van Valen, L. (1974). A natural model for the origin of some higher taxa. *Journal of Herpetology*, **8**, 109–21.

References to Chapter 2

1. Gould, S. J. (1977). Eternal metaphors of palaeontology. In *Patterns of Evolution*, ed. A. Hallam. Amsterdam: Elsevier.
2. Darwin, C. (1859). *On the Origin of Species by Means of Natural Selection, or the Preservation of Favoured Races in the Struggle for Life.* London: John Murray.
3. Bowler, P. J. (1983). *The Eclipse of Darwinism: Anti-Darwinian Evolution Theories in the Decades around 1900.* Baltimore, MD: Johns Hopkins University Press.
4. Bateson, W. (1894). *Materials for the Study of Variation, Treated with Especial Regard to Discontinuity in the Origin of Species.* London: Macmillan.
5. De Vries, H. (1910). *The Mutation Theory: Experiments and Observations on the Origin of Species in the Vegetable Kingdom.* (Translated by J. B. Farmer and A. D. Darbyshire; two volumes – II published in 1911.) London: Kegan Paul, Trench, Trübner & Co.
6. Goldschmidt, R. (1940). *The Material Basis of Evolution.* New Haven, CT: Yale University Press.
7. Kimura, M. (1983). *The Neutral Theory of Molecular Evolution.* Cambridge: Cambridge University Press.
8. Dover, G. A. (1982). Molecular drive: a cohesive mode of species evolution. *Nature*, **299**, 111–17.
9. Stanley, S. M. (1979). *Macroevolution: Pattern and Process.* San Francisco: Freeman.

10. Gould, S. J. and Lewontin, R. C. (1979). The Spandrels of San Marco and the Panglossian paradigm: a critique of the adaptationist programme. *Proceedings of the Royal Society of London Series B*, **205**, 581–98.

11. Alberch, P. (1980). Ontogenesis and morphological diversification. *American Zoologist*, **20**, 653–67.

12. Gould, S. J. (1989). A developmental constraint in *Cerion*, with comments on the definition and interpretation of constraint in evolution. *Evolution*, **43**, 516–39.

13. Arthur, W. (2001). Developmental drive: an important determinant of the direction of phenotypic evolution. *Evolution and Development*, **3**, 271–8.

14. Von Baer, K. E. (1828). *Über Entwicklungsgeschichte der Tiere: Beobachtung und Reflexion*. Königsberg: Bornträger.

15. Panchen, A. L. (2001). Étienne Geoffroy St.-Hilaire: father of 'evo-devo'? *Evolution and Development*, **3**, 41–46.

16. Haeckel, E. (1866). *Generelle Morphologie der Organismen*. Berlin: Georg Reimer.

17. Sander, K. (2002). Ernst Haeckel's ontogenetic recapitulation: irritation and incentive from 1866 to our time. *Annals of Anatomy*, **184**, 523–33.

18. Haeckel, E. (1896). *The Evolution of Man: A Popular Exposition of the Principal Points of Human Ontogeny and Phylogeny*. New York: Appleton.

19. Thompson, D'A. W. (1917). *On Growth and Form*. Cambridge: Cambridge University Press.

20. Huxley, J. S. (1932). *Problems of Relative Growth*. London: Methuen.

21. De Beer, G. R. (1940). *Embryos and Ancestors*. Oxford: Clarendon Press.

22. Schmalhausen, I. I. (1949). *Factors of Evolution: The Theory of Stabilizing Selection*. Philadelphia, PA: Blakiston.

23. Waddington, C. H. (1957). *The Strategy of the Genes*. London: Allen & Unwin.

24. Waddington, C. H. (1975). *The Evolution of an Evolutionist*. Edinburgh: Edinburgh University Press.

25. Fisher, R. A. (1930). *The Genetical Theory of Natural Selection*. Oxford: Clarendon Press.

26. Haldane, J. B. S. (1932). *The Causes of Evolution*. London: Longman.

27. Wright, S. (1931). Evolution in Mendelian populations. *Genetics*, **16**, 97–159.

References to Chapter 3

1. Darwin, C. (1859). *On the Origin of Species by Means of Natural Selection, or the Preservation of Favoured Races in the Struggle for Life.* London: John Murray.
2. Pasteur, L. (1854). (Address given 7 December: see Vallery-Radot, R. 1900, *La Vie de Pasteur*, chapter 4.)
3. Kimura, M. (1979). The neutral theory of molecular evolution. *Scientific American*, **241(5)**, 94–104.
4. Garcia-Bellido, A., Lawrence, P. A. & Morata, G. (1979). Compartments in animal development. *Scientific American*, **241(1)**, 90–8.
5. Dobzhansky, T. (1951). *Genetics and the Origin of Species*, third edition. New York: Columbia University Press.
6. Mayr, E. (1942). *Systematics and the Origin of Species.* New York: Columbia University Press.
7. Ford, E. B. (1971). *Ecological Genetics*, third edition. London: Chapman & Hall.
8. Simpson, G. G. (1944). *Tempo and Mode in Evolution.* New York: Columbia University Press.

References to Chapter 4

1. Spemann, H. (1938). *Embryonic Development and Induction.* New Haven, CT: Yale University Press.
2. Lawrence, P. A. (1992). *The Making of a Fly: The Genetics of Animal Design.* Oxford: Blackwell.
3. Meinhardt, H. (1982). *Models of Biological Pattern Formation.* London: Academic Press.
4. Gordon, R. (1999). *The Hierarchical Genome and Differentiation Waves: Novel Unification of Development Genetics and Evolution.* (Two volumes.) London: Imperial College Press.
5. Gilbert, S. (2000). *Developmental Biology*, sixth edition. Sunderland, MA: Sinauer.
6. Wolpert, L., Beddington, R., Jessell, T., Lawrence, P., Meyerowitz, E. and Smith, J. (2002). *Principles of Development*, second edition. Oxford: Oxford University Press.
7. Arthur, W. (1987). *Theories of Life: Darwin, Mendel and Beyond.* Hardmondsworth: Penguin.
8. Medawar, P. B. (1967). *The Art of the Soluble.* London: Methuen.

References to Chapter 5

1. Clarke, B. (1975). The causes of biological diversity. *Scientific American*, **233**(2), 50–60.
2. Seilacher, A. (1989). Vendozoa: organismic construction in the Proterozoic biosphere. *Lethaia*, **22**, 229–39.
3. Eldredge, N. and Gould, S. J. (1972). Punctuated equilibria: an alternative to phyletic gradualism. In *Models in Paleobiology*, ed. T. J. M. Schopf. San Francisco: Freeman.
4. Mundel, P. (1979). The centipedes (Chilopoda) of the Mazon Creek. In *Mazon Creek Fossils*, ed. M. H. Nitecki, pp. 361–78. New York: Academic Press.

References to Chapter 6

1. Kuhn, T. S. (1970). *The Structure of Scientific Revolutions*, second edition. Chicago: University of Chicago Press.
2. Koestler, A. (1967). *The Ghost in the Machine*. London: Hutchinson.
3. Arthur, W. (1997). *The Origin of Animal Body Plans: A Study in Evolutionary Developmental Biology*. Cambridge: Cambridge University Press.
4. Gould, S. J. (1977). *Ontogeny and Phylogeny*. Cambridge, MA: Harvard University Press.
5. Fisher, R. A. (1930). *The Genetical Theory of Natural Selection*. Oxford: Clarendon Press.
6. Hennig, W. (1981). *Insect Phylogeny*. Chichester: Wiley.
7. Whyte, L. L. (1965). *Internal Factors in Evolution*. London: Tavistock Publications.
8. Yampolsky, L. Y. and Stoltzfus, A. (2001). Bias in the introduction of variation as an orienting factor in evolution. *Evolution and Development*, **3**, 73–83.

References to Chapter 7

1. Dawkins, R. (1976). *The Selfish Gene*. Oxford: Oxford University Press.
2. Gould, S. J. (1977). *Ontogeny and Phylogeny*. Cambridge, MA: Harvard University Press.
3. McKinney, M. L. and McNamara, K. J. (1991). *Heterochrony: The Evolution of Ontogeny*. New York: Plenum Press.

4. Raff, R. A. (1996). *The Shape of Life: Genes, Development and the Evolution of Animal Form*. Chicago: Chicago University Press.
5. Arthur, W. (2000). The concept of developmental reprogramming and the quest for an inclusive theory of evolutionary mechanisms. *Evolution and Development*, **2**, 49–57.

References to Chapter 8

1. Gould, S. J. (1989). *Wonderful Life: The Burgess Shale and the Nature of History*. London: Hutchinson Radius.
2. Wallace, A. R. (1870). *Contributions to the Theory of Natural Selection: A Series of Essays*. London: Macmillan.
3. Klingenberg, C. P. (2002). Morphometrics and the role of the phenotype in studies of the evolution of developmental mechanisms. *Gene*, **287**, 3–10.
4. Wallace, A. R. (1897). *Darwinism: An Exposition of the Theory of Natural Selection, with some of its Applications*. London: Macmillan.

References to Chapter 9

1. Goldschmidt, R. (1940). *The Material Basis of Evolution*. New Haven: Yale University Press.
2. Dawkins, R. (1986). *The Blind Watchmaker*. London: Longman.
3. Ford, E. B. (1971). *Ecological Genetics*, third edition. London: Chapman & Hall.
4. Fisher, R. A. (1930). *The Genetical Theory of Natural Selection*. Oxford: Clarendon Press.
5. Arthur, W. (2001). Developmental drive: an important determinant of the direction of phenotypic evolution. *Evolution and Development*, **3**, 271–8.
6. Yampolsky, L. Y. and Stoltzfus, A. (2001). Bias in the introduction of variation as an orienting factor in evolution. *Evolution and Development*, **3**, 73–83.
7. Cohen, J. and Stewart, I. (2002). *Evolving the Alien: The Science of Extraterrestial Life*. London: Ebury Press.

References to Chapter 10

1. Darwin, C. (1859). *On the Origin of Species by Means of Natural Selection, or the Preservation of Favoured Races in the Struggle for Life*. London: John Murray.

2. Fisher, R. A. (1930). *The Genetical Theory of Natural Selection*. Oxford: Clarendon Press.
3. Whyte, L. L. (1965). *Internal Factors in Evolution*. London: Tavistock Publications.
4. Wagner, G. P. and Schwenk, K. (2000). Evolutionarily stable configurations: functional integration and the evolution of phenotypic stability. *Evolutionary Biology*, **31**, edited by Max K. Hecht *et al*. New York: Kluwer Academic/Plenum Publishers.
5. Arthur, W. (1997). *The Origin of Animal Body Plans: A Study in Evolutionary Developmental Biology*. Cambridge: Cambridge University Press.

References to Chapter 11

1. Von Baer, K. E. (1828). *Über Entwicklungsgeschichte der Tiere: Beobachtung und Reflexion*. Königsberg: Bornträger.
2. Darwin, C. (1859). *On the Origin of Species by Means of Natural Selection, or the Preservation of Favoured Races in the Struggle for Life*. London: John Murray.
3. Richardson, M. K. and Keuck, G. (2001). A question of intent: when is a 'schematic' illustration a fraud? *Nature*, **410**, 144.
4. Wray, G. A. and Raff, R. A. (1989). Evolutionary modification of cell lineage in the direct-developing sea urchin *Heliocidaris erythrogramma*. *Developmental Biology*, **132**, 458–70.
5. Panchen, A. L. (1992). *Classification, Evolution and the Nature of Biology*. Cambridge: Cambridge University Press.

References to Chapter 12

1. Waddington, C. H. (1956). Genetic assimilation of the bithorax phenotype. *Evolution*, **10**, 1–13.
2. Schlichting, C. D. and Pigliucci, M. (1998). *Phenotypic Evolution: A Reaction Norm Perspective*. Sunderland, MA: Sinauer.
3. Brakefield, P. M., Gates, J., Keys, D., Kesbeke, F., Wijngaarden, P. J., Monteiro, A. *et al*. (1996). Development, plasticity and evolution of butterfly eyespot patterns. *Nature*, **384**, 236–42.

References to Chapter 13

1. Gerhart, J. and Kirschner, M. (1997). *Cells, Embryos and Evolution: Toward a Cellular and Developmental Understanding of Phenotypic Variation and Evolutionary Adaptability.* Malden, MA: Blackwell.
2. Raff, R. A. (1996). *The Shape of Life: Genes, Development and the Evolution of Animal Form.* Chicago: Chicago University Press.
3. Ohno, S. (1970). *Evolution by Gene Duplication.* New York: Springer-Verlag.
4. Dawkins, R. (1986). *The Blind Watchmaker.* London: Longman.
5. Rutherford, S. L. and Lindquist, S. (1998). Hsp90 as a capacitor for morphological evolution. *Nature,* **396**, 336–42.

References to Chapter 14

1. Arthur, W. (1997). *The Origin of Animal Body Plans: A Study in Evolutionary Developmental Biology.* Cambridge: Cambridge University Press.
2. Sneath, P. H. A. and Sokal, R. R. (1973). *Numerical Taxonomy: The Principles and Practice of Numerical Classification.* San Francisco: Freeman.
3. Hennig, W. (1966). *Phylogenetic Systematics.* Urbana: University of Illinois Press.
4. Aguinaldo, A. M. A., Turbeville, J. M., Linford, L. S., Rivera, M. C., Garey, J. R., Raff, R. A. and Lake, J. A. (1997). Evidence for a clade of nematodes, arthropods and other moulting animals. *Nature,* **387**, 489–93.
5. De Rosa, R., Grenier, J. K., Andreevas, T., Cook, C. E., Adoutte, A., Akam, M. *et al.* (1999). Hox genes in brachiopods and priapulids and protostome evolution. *Nature,* **399**, 772–6.

References to Chapter 15

1. Scott, M. P. and Weiner, A. J. (1984). Structural relationships among genes that control development: sequence homology between the *Antennapedia, Ultrabithorax* and *fushi tarazu* loci of *Drosophila. Proceedings of the National Academy of Sciences of the USA,* **81**, 4115–19.
2. McGinnis, W., Garber, R. L., Wirz, J., Kuroiwa, A. and Gehring, W. J. (1984). A homologous protein-coding sequence in *Drosophila* homeotic genes and its conservation in other metazoans. *Cell,* **37**, 403–8.

3. Lawrence, P. A. (1992). *The Making of a Fly: The Genetics of Animal Design*. Oxford: Blackwell.

4. Davis, G. K. and Patel, N. H. (1999). The origin and evolution of segmentation. *Trends in Genetics*, **15**, M68-M72 (Millennium issue).

5. Clark, R. B. (1964). *Dynamics in Metazoan Evolution: The Origin of the Coelom and Segments*. Oxford: Clarendon Press.

6. Diaz-Benjumea, F. J., Cohen, B. and Cohen, S. M. (1994). Cell interaction between compartments establishes the proximal-distal axis of *Drosophila* legs. *Nature*, **372**, 175–9.

7. Panganiban, G., Irvine, S. M., Lowe, C., Roehl, H., Corley, L. S., Sherbon, B. *et al.* (1997). The origin and evolution of animal appendages. *Proceedings of the National Academy of Sciences of the USA*, **94**, 5162–6.

References to Chapter 16

1. Fisher, R. A. (1930). *The Genetical Theory of Natural Selection*. Oxford: Clarendon Press.

2. Dawkins, R. (1986). *The Blind Watchmaker*. London: Longman.

3. Gould, S. J. (1977). Eternal metaphors of palaeontology. In *Patterns of Evolution*, ed. A. Hallam. Amsterdam: Elsevier.

4. Fusco, G. (2001). How many processes are responsible for phenotypic evolution? *Evolution and Development*, **3**, 279–86.

5. Minelli, A. (2000). Limbs and tail as evolutionarily diverging duplicates of the main body axis. *Evolution and Development*, **2**, 157–65.

References to Chapter 17

1. Gould, S. J. and Eldredge, N. (1986). Punctuated equilibrium at the third stage. *Systematic Zoology*, **35**, 143–8.

2. McNamara, K. J. (1997). *Shapes of Time: The Evolution of Growth and Development*. Baltimore: Johns Hopkins University Press.

3. Cheverud, J. M. (1984). Quantitative genetics and developmental constraints on evolution by selection. *Journal of Theoretical Biology*, **110**, 155–71.

4. Goldschmidt, R. (1940). *The Material Basis of Evolution*. New Haven: Yale University Press.

5. Mendel, G. (1866). Versuche über Pflanzenhybriden. *Verhandlungen des naturforschenden Vereines in Brünn, Bd. IV für das Jahr 1865*, Abhandlungen, 3–47.

Index

Page numbers in *italics* refer to figures.

adaptation 36, *118*, 117–27
Alberch, P. 16
anatomy 166
ancestor 6, 18, 152
annelid 170
arthropod 170
assimilation, genetic 19, 146
axis, body 133, 155, 179

Bateson, W. 13, 106
bias, developmental *12*, 16, 101–2, 103–4,
    105–16, 195, 201
bicoid 48
Bilateria *173*
bithorax 145
bivalve 184
body plan 31–2, 46–7, 62, 65, 152
Bonner, J. T. 1
Brakefield, P. 147
breeding 205
butterfly 147–8

Cambrian
    explosion 61
    Period 58
cascade 50, 51, 85, 176, 187, 188
cell number 1, 7
centipede *59*, 63–4, *177*, 206
character 167–70
Cheverud, J. 205
chick *41*
chimpanzee 163–4, *165*
chordate 170
cladistics 167–70
Clark, R. B. 182
Clarke, B. 56
coadaptation 36, 117–27, 195
computer 169
constraint, developmental 16, 74, 115, 205
contingency 88, 102, 103–4, 113–16

convergence 169–70, 184–5
co-option 186, 198
covariation *99*
creationism 70

Darwin, C. 9–13, 26–39, 65, 70, 106, 119,
    134
Dawkins, R. 76, 109, 192–3
de Beer, G. 19
de Vries, H. 14, 110
Deuterostomia *173*
digit 133–4
disc, imaginal 45–6, 136
distal-less 176, *177*
distribution, frequency 90–1
DNA 175–6
Dobzhansky, T. 34
Dover, G. 15
drift, genetic 14–15, 38
drive
    developmental 16, 116
    molecular 15
duplication and divergence 154–6, 189–90,
    197

Ecdysozoa *173*
echinoderm 170
ecology, behavioural 23–4
Ediacaran fauna 59
egg 47–8
egg timer 135
Eldredge, N. 62
engrailed 49, 176, *177*, 186–7
environment 122–7, 140–51, 196
epigenesis 85
evolution, direction of 37–9, 88–116
evolvability 153–8
explosion, evolutionary 61–2
eye 182
eyespot 147–8

Fabricius, H. 42
faith 209
Fisher, R. A. 34, 73, 106, 109, 110, 119
fitness 95, 96, 122–7
    profile 126, 125–7, 196
Ford, E. B. 34, 109, 110
fossil 24, 55, 60, 57–65
frog 41
fruitfly 41, 43, 145–6, 179, 188
fundamentalism 210
Fusco, G. 196

gene
    duplication 15
    expression 49–50, 177, 177
    names 48
    number 7
    selfish 76–8
    switching 48–50, 176
generalized 156–8
genetics
    developmental 24
    ecological 23, 27
    population 22–3, 159
    quantitative 23, 204–5
genome 7
Geoffroy St Hilaire, E. 17
Goldschmidt, R. B. 14, 106, 108, 208
Gordon, R. 51
gorilla 163–4, 165
Gould, S. J. 11, 16, 25, 62, 72, 88
gradient 48
growth, correlation of 31

Haeckel, E. 17–18, 135
Haldane, J. B. S. 34
hemichordate 170
Hennig, W. 74, 167
heritability 205
heterochrony 81–2, 195, 204
heterometry 83
heterotopy 82, 195
heterotypy 83
hierarchy 50
history
    of developmental biology 41–3
    of evolutionary biology 9
homeobox 175–8
homeodomain 176
homeosis 145, 175
Hominidae 163–4

homology 24, 152–3, 183–4
Hox gene 155–6
human 163–4, 165
hunchback 48
Huxley, J. 19

insect 177, 188

Kimura, M. 14–15, 29
Klingenberg, C. 93
Koestler, A. 71
Krüppel 49
Kuhn, T. 69

Lamarck, J. 11, 19
landscape, adaptive 94, 93–102, 103–4,
    107–15, 207–8
larva 129, 136
Lawrence, P. 47, 179
Lewis, E. 43
Lewontin, R. C. 16, 25
life cycle 1–3, 6–8, 32
limb 154–5, 183–4, 187–9
Lophotrochozoa 173

macroevolution 62
Malpighi, M. 42
mammal 98, 128–9
Mayr, E. 34
McKinney, M. 81
McNamara, K. 81, 204
Medawar, P. 53
megaevolution 62
Meinhardt, H. 51
Mendel, G. 33, 210
metamorphosis 45, 45–6
method, scientific 68–9, 209
microevolution 62
model system 40–3
modern synthesis 34–9, 73–4, 191–4,
    199–200
modularity 153–6, 188, 197
mollusc 170
monster, hopeful 108
mouse 41
mutation 35–8, 79, 105–16, 132–3
mutationism 13–14

natural history 159
natural philosophy 159
nematode 170, 171

nemertine 170, *171*
neo-Darwinism 20, 34–9, 191–4
novelty, evolutionary 78–82

organizer 42
orthogenesis 13, 74, 121
outgroup 168, *168*

palaeontology 159
Panchen, A. 139
pan-externalism 23, 36, 119
pan-selectionism 23, 38, 98, 119
parsimony 182
Pasteur, L. 26
pathway, developmental 49–50
Phanerozoic era 61
phenetics 166–7
phenocopy 145–6
phylogenetics 20–2
phylogeny *173*, 159–74, 197
Pigliucci, M. 147
placenta 129
plasticity 144–51
polyphenism 148
Popper, K. 69
population 79–80
primordium 153
Protostomia *173*
punctuated equilibrium 62, 202

Raff, R. 137
reaction norm *150*, 146–51, 196
recaptitulation 17–18
religion 209
reprogramming, developmental *84*, 84–7,
        114–16, 151, 195
revolution, scientific 69–74
Richardson, M. 135
Roux, W. 42
rudiment 136

saltation 109
scenario, adaptive 140
Schlichting, C. 147
Schmalhausen, I. 19
Schwenk, K. 124
segmentation 46, 179, *181*, 185–7
Seilacher, D. 60
selection
        external 123–7, 128

internal 121–7, 128, 195
kin 77
natural 27–8, 32–3, 70–3, 80, 97, 103–4,
        113–16, 201
species 15
shape 142
Simpson, G. G. 34, 62
snail 133, *141*, 140–5
Sneath, P. 166
sociobiology 23
Sokal, R. 166
specialized *157*, 156–8
Spemann, H. 42
Stanley, S. 15
starfish 188
Stoltzfus, A. 74, 116
strawman 120, 192
symmetry 178
systematics 167

tagma 179
tardigrade 170, *171*
thale cress *41*
theory, general 159, 209
Thompson, D'A. 19
trait 52
transplant experiment 143–5
tree, evolutionary *4*, 3–6, *162*, *168*, 159–74

van Valen, L. 7
variation 10–11, 14, *91*, 88–93, *96*, 105, 158,
        200
Vendozoa 59, *60*
vertebrate *59*, 62–3
von Baer, K. E. 16–17

Waddington, C. H. 19–20, 145
Wagner, G. 124
Wallace, A. R. 12–13, 89–90, 97, 100, 106
water bear 170, *171*
whale 104
Whyte, L. L. 74, 121
worm *41*, 170, *171*
Wray, G. 137
Wright, S. 34, 93

Yampolsky, L. 74, 116
yolk 135

zebrafish *41*